Conspiracy

ALFRED ADAMS

Conspiracy
Copyright © 2023 by Alfred Adams

ISBN:
Paperback: 979-8-9878888-0-3
Hardback: 979-8-9878888-2-7
e-book: 979-8-9878888-1-0

All rights reserved. No part of this publication may be reproduced, distributed, or transmitted in any form or by any means, including photocopying, recording, or other electronic or mechanical methods, without the prior written permission of the publisher, except in the case brief quotations embodied in critical reviews and other noncommercial uses permitted by copyright law.

The views expressed in this book are solely those of the author and do not necessarily reflect the views of the publisher, and the publisher hereby disclaims any responsibility for them.

Contets

INTRODUCTION v
Chapter 1: NEW YORK CITY 1
Chapter 2: GLENDALE, CALIFORNIA 10
Chapter 3: U. S. MARINE CORPS 21
Chapter 4: RETURNING HOME 41
Chapter 5: JACQUE COUSTEAU 50
Chapter 6: MY CAREER WITH THE FAA 66
Chapter 7: LEAVING THE FAA 92
Chapter 8: ARRIVA AIR INTERNATIONAL 111
Chapter 9: SAUDI ARABIA 116
Chapter 10: STONE AIR AVIATION 133
Chapter 11: ARIZONA DEPARTMENT OF TRANSPORTATION 137
EPILOGUE 143

INTRODUCTION

The cover is a type of before and after photos and classified by the City of Phoenix, as the worst air disaster in the history of Sky Harbor Airport. There were eight passengers and crew on board, and everyone got off without so much as a scratch.

The final NTSB and FAA investigative reports of this accident revealed the FAA removed two key witness statements from the final accident report that would have shown the cause of this accident to be that both sets of brakes, pilot and co-pilot's brake pedals froze in the full up position and unusable. The brake pedals are the only method of applying normal or emergency brakes.

With the removal of this crucial evidence, the FAA was able to blame the accident on the pilot because he did not go through the procedure of using emergency braking, even though the emergency brakes were inoperable. Without any brakes to stop the aircraft, it went through the perimeter fence, crossed 24th street, a busy thoroughfare and crashed through a six-foot concrete block wall, bursting into flames.

Another lie by the FAA is when they blamed the brake failure on the Anti-Skid System. The manufacturer of the Anti-Brake System tested the entire unit, and it worked fine with no malfunctions. The FAA changed the final accident report by removing two witness statements. Removing the two witness statements eliminated the fact the brake pedals were frozen in the full up position making it impossible to operate either the normal or emergency brake systems. The FAA made it appear as if the pilot didn't know how to use the emergency brake thereby justifying taking enforcement action against the pilot when they attempted to suspend his pilot certificate for forty-five days. The pilot took the FAA to task and appealed the case to the NTSB Hearing Judge and won after all the facts were brought out.

This book reveals the FAA's Conspiracy against an FAA employee that began when an FAA Air Traffic Controller demanded the pilot of a Mooney 231, to execute an unsafe maneuver causing the airplane to crash killing both the brother of the FAA employee and the brother's son-in-law, Vince.

"Conspiracy" is an autobiography of the accomplishments of the pilot and what he did in an attempt to escape the FAA's Vendetta. However, the FAA's vendetta followed him throughout his aviation career.

© Copyright

Chapter 1

NEW YORK CITY

New York City is made up of five Burroughs, Manhattan, Brooklyn, Bronx, Staten Island, and Queens. I grew up in Springfield Gardens located in the Borough of Queens on Long Island at the east edge of New York City's line just before Nassau County. You enter Nassau County which is outside and east of the city. The Southern State Parkway is the main highway and the boundary line between New York City and Nassau County or Long Island.

Alongside the Parkway were small ponds or lakes all in a row each connected by a small stream. These ponds or lakes, whatever you wanted to call them weren't very clean, and the color of the water was brown. During the summer months, we would go swimming in one of these lakes until my mother came down to watch us. She was disgusted when she saw garbage flowing into our swimming hole and immediately took me home.

The house where I grew up was on 222nd Street between 131st and 132nd Avenue in Springfield Gardens. The property was only fifty feet wide by one hundred feet deep. Even though it was very narrow, it still had a driveway down the side to the garage. So, you can imagine the size of the house.

My father was a New York City Police Officer at the 103rd Precinct in Jamaica. In our garage, my dad had set up a repair shop where he worked on the personal cars of Police Officers from the 103rd and 105th Precincts. My father also had a third job as a mechanic on the Green Bus Line.

The rear yard contained a small fish pond which held three or four goldfish. It did not have any filtering system, so we had to change the water every so often when it got black. During the winter months the pond would freeze solid, but somehow the fish survived until the summer when the ice melted. We never fed the fish. There were so many mosquitoes flying around that they had plenty to eat.

The house was very tiny and had only two bedrooms, one for my folks and the other for my brother and me. Eventually, my parent's built another bedroom onto the rear of the house over the kitchen, and that became their bedroom.

My mother took care of an older woman we called Mother Johnson. She occupied my parent's old bedroom. Being an invalid, she was only able to get from her room to the bathroom by holding onto her wooden wheelchair. She would support herself by holding onto the arms of the wheelchair and push it in front of her into the bathroom. Sometimes she would be naked when she walked from the bedroom to the bathroom. As a kid, it was horrible for my brother and me to see this.

The front of our house had a screened in porch where my friends, Bubby Watson, Ron Miller, Eddy Lurk, Richard Silva and I would sleep in the summer to take advantage of the cooler outside air. The one thing we couldn't beat was the humidity. It was awful and kept us awake until all hours of the night. Air conditioning was unheard of in the forties and fifties.

The guys I hung around with always tried to look and act older than we were so, we took up smoking cigarettes because we thought it would be very cool. I was eight years old when I began my smoking career. Buying them was no problem because we always told the clerk they were for our parents. The clerk didn't care; they just wanted the business.

Two blocks from my house was a huge vacant piece of land that had what we called the jungle. It was overgrown with trees and bushes which had to be chopped down to gain access to the center where we built the "Fort." We dug two holes; one was the main room, which measured about ten by ten-foot square and six foot deep and the other was the entrance. We cover the hole with boards and sheets of plywood, and on top we planted bushes. To prevent the roof from collapsing, we installed a 4 X 4 wooden support attached to a cross beam in the center of the ceiling.

Saturday night was the time when my mother gathered all our dirty clothes for the laundry. She was cleaning out the pockets of my pants and came across some prophylactics. Naturally, she was shocked because I was only 12 years old. She told me she was going to tell my father, but when he confronted me, he only said he was glad I was taking precautions.

In 1951, I graduated from Grammar School and surprised everyone. That September I started my short stay in High School. During my high school years, I was becoming an uncontrollable hoodlum hanging out with a huge street gang called Scorpions. Our numbers were so large that a rival gang came to Springfield Gardens to fight. We told them we would meet them at an abandon baseball field. They showed up in eight cars, but when they realized we outnumbered them by about four to one, they took off without as much as a "hello." The baseball field was full of our gang members, almost shoulder to shoulder. I couldn't even guess how many of us were there.

Every so often Bazars would come and set up tents on vacant land. There would be games in these tents where people would bet on numbers, and the attendant would spin a wheel. Three- inch metal spokes stuck out from the outside perimeter of the wheel creating a clicking noise as they slapped against a wooden pointer. If the pointer stopped on their number, they won a price. The prizes were bottles of liquor.

One foggy night, because of the limited visibility, no one saw was able to see us when we crawled under one of the tents and stole four bottles of liquor. The fog hampered the security guard's vision and he never even knew we were there.

I took my bottle to school removed the cap and stuck it inside my jacket. I then stuck a long straw inside the bottle with the other end stuck inside the collar of my coat. When I wanted a drink, I would bring the straw out of my jacket and sip the liquor. As I sat in class drinking this liquor, it didn't take me long until I was very drunk.

In the lunchroom, a teacher smelled the alcohol and was about to turn me over to security, when I ran from the school and played hooky the rest of the day. I fell asleep lying in the middle of the sidewalk in front of a candy store across the street from the school.

Tony Fortunato, a high school friend, came along, woke me up and we got out of the area. We walked along the main street of Linden Boulevard looking in the window of parked cars to see if anyone had left their keys. We were walking through a parking lot when we spotted a key ring containing keys hanging from the dashboard of a big Packard.

I got in the driver's seat and started inserting each key into the ignition. Finally, one worked, and I started the car, drove it out of the parking lot and onto the street. Just as we were leaving the parking lot, the owner of the car saw us and came running out of a store. Still a little drunk from the liquor, I stopped the car wanting to see how fast it would accelerate from a dead stop. The owner was still running, when Tony started yelling at me to get the hell out of there. When the owner got about twenty feet from the rear of the car, I slammed my foot down hard on the accelerator. The car lurched ahead and sped off down the street leaving the owner screaming and shaking his fist.

We left the area believing the owner would be calling the police immediately and we didn't want to be anywhere near there. We drove east toward the county line so that we wouldn't be that far from home when we abandoned the car. We changed drivers until we were tired of it and then parked the car on a side street and walked out of the area.

During summer vacation, when there was no school, the guys I hung around with on a regular basis, would take our bikes and ride to Rockaway Beach, a distance of about 20 miles. The area of Rockaway Beach where we liked to swim was in an inlet where the water was calm and not as rough as the ocean. The inlet had two moored rafts made of wood and held afloat by eight fifty gallon drums secured underneath. These rafts were about fifteen feet square and two feet above the water. They were perfect for just lounging around on the surface or diving from them.

We had to be careful walking in the water because Horse Shoe Crabs were lying on the bottom in the sand. These crabs had a triangular shaped bone tail approximately seven inches long, and rumor had it, if they were asleep, their tail would stand straight up. If you weren't careful, you could step on the tail, and it would go straight up through your foot. You had to remember to drag your feet along the sandy bottom. These crabs were downright ugly. They looked like half a basketball and were gray. When you turned them over, there were many crab pinchers.

During the winter months and off-season for tourists, we would ride to the beach and break into the vacant houses. There was nothing to steal, but we had fun just rummaging through all those houses.

A friend and I drove a stolen car to Tony's house and as we approached we slowed and came to a stop around the corner. We got out leaving the car running because we had the ignition hot-wired. As I walked up to Tony's house, there were two policemen at his front door. Tony signaled me from the front window to get out of there. I turned and walked back around the corner but didn't go to the car we had waiting. I looked back, and one policeman walked to the corner and was watching me. I walked past the car, and the cop went back to Tony's house. I went back to the car got in, made a "U" turn and got out of there.

Growing up, many times I was questioned by the New York City's Police from the 105th Prescient for doing something out of the ordinary. My father was a Police Officer in the 103rd Prescient in Jamaica, which was a suburb of New York City. Every time a Police Officer from the 105th Prescient stopped and questioned me, they would call my father and tell him what I was doing and make it sound as if I was in deep trouble. I never found out if they were kidding, but in any case, I was in trouble with my father when he got home.

He was a huge man but only stood five foot 10 inches tall. He was very broad, but it was not fat. He didn't know his strength. I saw him take a complete automobile engine and lift it into the back of a pickup truck by himself. My mother would always be concerned when he spanked my brother or me because of his strength.

My older brother, George or Buddy as we called him, was older than me by three and a half years. We both went to the same High School. He was always the honor student and could do no wrong by the school faculty. When they heard his little brother was coming to the High School, they praised their luck. Boy, did I show them how wrong they were?

My brother and I were total enemies and would fight every chance we got. He was much bigger than me, so he always won. Even though we hated one another, I had a lot of respect for him because when he graduated from high school, he joined the Navy. On his own, he went to classes and specialized schools qualifying him to enter the U.S. Naval Academy at Annapolis, Maryland. Many of the students in that university got there through congressional appointments, but he made it on his own. The son of a New York City Cop going to Annapolis Naval Academy was something special. My mother and father were proud of him and justifiably so.

When I was about fourteen years old, he was home for the Christmas Holidays. He asked me if I wanted to go into Brooklyn drinking with him and his friend Johnny China. Naturally, I said yes, and from that moment on, we became the best of friends.

By the time I entered high school, I was becoming a Supreme hoodlum. Four of my friends and I would cut school and roam up and down the streets looking into automobiles hoping someone was stupid enough to leave their keys in the ignition. When they did, we would steal that car and go joy riding. We kept score on a total number of cars we took. To my recollection, I had twenty- three to my record.

One day we took a 1936 Buick Touring Car. The car is a four-door convertible, with manual shift and it looked as if it was right out of a gangster movie, but the paint on the car was a light puke green. It was a great car, and we liked it so much we had an extra set of keys made. Then, before stealing the car again, we would put a mark on the fuel gauge showing the position of the gas before we left. Just before returning the car to the exact location where we stole it from, we would fill the gas tank to the same level as when we left. We did this for two weeks until we got tired of driving it. Stealing cars was how I learned to drive.

While my brother was in his first year at Annapolis, he wasn't allowed to have a car, so he left it at home. When my mother and father would go out for the evening, my friends and I would go joy riding in his car because when the opportunity arose, I had extra keys made.

One night my friends and I had my brother's car, and we were touring all over Springfield Gardens when the battery went dead. Fortunately, the car was stick shift, so when the engine stalled, which happened quite frequently being an unskilled driver, my friends would get out and push it. I would put it in gear, pop the clutch and I would go speeding away. They would run after me until I stopped and everyone piled in the back, and we'd be off again. The car was a Pontiac convertible, so it was easy for everyone to jump into the rear seat.

In High School, I was on the dean's list, but this was not a list of good students. Anytime there was trouble in the high school I would be the one to be questioned. One day when the weather was cold, two students, whom I didn't know, built a small fire between two teacher's cars to keep warm. One of the cars caught fire. The fire engines showed up and put the fire out. Many people saw me in the area, so everyone thought I was to blame. Nothing could be proven, but I was questioned extensively by the police.

My life went this way until I entered my Junior (3rd) year of High School. By that time I had had enough of school and got permission to quit. I got a job as a stock clerk in O'hrback's department store in downtown Manhattan.

My commute to Manhattan was to take a bus from my home, travel three miles to Jamaica and then switch to the subway train for forty-five minutes to 34th Street and Broadway. The subway cars were so crammed full of passengers, many times I didn't have to hold onto something because the force of everyone leaning against me kept me from falling over. If I was able to get a seat, the noise from the train was so loud and constant it would hypnotize me, and I would fall asleep.

The subway system was dangerous. I was surprised more people didn't lose their lives while waiting on the platform for their train. The people commuting would stand on the platform with their feet a mere six inches from the edge as the train pulled in to the station doing about forty miles per hour. People in the rear would start positioning themselves to be in front of the door of the train when it stopped. Those in front would push rearward to keep from being forced onto the train tracks as the train arrived. As the train pulled into the station, everyone would jostle sideways up and down the platform trying to adjust their position so that when the door opened, it would be directly in front of them. It was the only way they could expect to get a seat. However, even this technique didn't work most of the time.

With my brother in the Naval Academy, every year he was able to get us tickets for the Army and Navy football game in Philadelphia, Pennsylvania. The game was in November, and my father, Uncle Tom and my cousin Tommy and I would drive down from New York. One year it was so cold when we arrived in Philadelphia, we all bought a bottle of liquor to try to keep warm. I didn't drink because I didn't like the taste. So, by the end of the game, I was the only person sober enough to drive home to New York City. Fortunately, stealing all those cars taught me how to drive. My father always thought I learned to drive using my brother's car. I was fourteen years old.

The basement of our house underwent remodeling by breaking down the front concrete block wall exposing a wall of dirt fifteen feet wide by seven feet high. We shoveled the dirt away and installed an antique wooden bar with enough room a person could serve from behind it. Under the bar, we placed a beer keg along with all the necessary plumbing.

Before we completed rebuilding the wall, my brother and I would set targets in the dirt and shoot them with .22 rifles. We also shot my 30/30 Winchester, and 20 gauge shotgun. I don't know how the neighbors kept from hearing the noise from these guns, but we never received a complaint.

At work, a friend sold me his .25 caliber semi-automatic pistol with pearl handles. It wasn't in good condition, but I bought it anyway. When I got home, I took into our basement and shot it into the target area (dirt). It worked just fine. From that day forward I carried a gun with me wherever I went.

When the basement was complete and the bar installed and working, my mother and father had parties inviting all their friends. The following day, after the party was over, we would invite my friends over, and we would share the beer left in the keg. The only demands were, no one was to leave the house, and everyone would stay the night.

FIGURE 1 - MYSELF, EDDIE LURK, RICHARD SILVA

During the summer of 1953, my parents and I made a trip to Los Angeles, California hoping and praying for possibly moving the following year.

In June of 1954, my father retired from the New York City, Police Department. I was fifteen and my brother still in the Navy and I had decided that the two of us would move to California when he got out of the Navy. However, before we moved, our parents had already decided to move to Los Angeles, California that year.

Before leaving, we had to sell our house in New York. The first person to look at the house bought it. When my father first showed the house to the couple who bought it, the husband threw his arms over the bar and said this is his place now. They sold the house at the asking price. In my heart to this day, this move saved my life because I was heading on a downhill slide

Conspiracy

toward bigger problems I couldn't stop. It was just a matter of time before I would be caught by the police doing something crazy and end up in prison.

I told my brother of dad's plans to move to California, and his reaction was very positive. He got some time off to help with the move and came home. He had an Oldsmobile which he and I drove while mom and dad towed a trailer behind their 1955 Buick Century. The trip was uneventful, and we arrived in Glendale, California five days after leaving New York City.

Chapter 2

GLENDALE, CALIFORNIA

Traveling from New York to California was uneventful. My mom and dad rented a small apartment in Glendale which is a quiet suburban community thirteen miles north of downtown Los Angeles. I immediately applied to and became employed at the O'hrback's Department store again, as a stock clerk.

My father leased a Richfield Gas Station and Auto Repair Shop on the corner of San Fernando Road and Chevy Chase Road in Glendale. I quit O'hrback's and worked for my father dispensing gas and lubricating customer's automobiles. The location of the gas station was in an unsavory neighborhood called Toonerville. We worked Monday thru Saturday and had Sunday off.

My father helped me buy my first car. It was a 1953 MG TD Roadster. MG stands for Morris Garage. It was a convertible with two doors and tan in color. The location of Morris Garage is in England, and the model designation of TD existed between the years 1950 and 1953. While the top was down, the cockpit had a tonneau cover protecting the seats from the sun. The spare tire was exposed and mounted on a bracket on the rear of the car just over the gas tank. It had a cover over the tire that was of the same design and color as the cockpit tonneau cover. The engine had four cylinders with a four-speed stick transmission. The seats were red leather with the red plaid tonneau cover. The gas gauge was a green light on the dashboard which when lit signified there were two gallons of gasoline left. However, on my car when the green light came on the engine quit almost immediately. I then made it a habit to top the fuel tank when the odometer hit two hundred miles.

Shortly after my father and I opened the gas station for business, an MG TC drove into the repair shop. Since it was the same make as my car, I went to the rear and introduced myself to Greg Knapp. We immediately became friends.

Greg asked my dad if he could determine why the front left wheel felt like it was wobbling. My dad said the pinion bearings were worn out and needed replacing. Greg left his car to be repaired and came to the front of the station where I was working. I found that Greg's full name was Gregory Cromwell Knapp. We became friends immediately and made arrangements to go to Hollywood that night in my MG. Hollywood Boulevard was the hangout for high school kids looking to meet. We had a great time and cemented a friendship that has lasted until this very day. We both have lived all over this world, and we've ended up living eight blocks from one another.

With Greg's MG up on jacks and wheels removed, I looked at the left front spindle (axle), and it appeared to have a crack at its base. I wiggled it, and it fell into my hand. It was fortunate Greg had come into the shop when he did before the front wheel came off. I'm sure the car would have gone out of control and crashed.

Greg and I became best of friends and went everywhere together. Since his MG was in the shop for repairs, we used my MG. We would drive to Hollywood Boulevard which was the hang out for high school kids meeting other kids. An MG was an unusual type of car, so it was a big hit in Hollywood.

Within six months the family moved to a larger apartment and in a much better neighborhood. My mother organized a housewarming party and invited all our relatives and friends.

Greg and I had just returned from shooting in the desert and were in my bedroom preparing to clean my .22 semi-automatic rifle. The rifle had a loading tube just below the barrel. I made sure all cartridges were removed by taking the insert from the tube and placing it on top of the dresser.

I then took the rifle and tapped the open end of the tube on the floor to dislodge any cartridges. None came out.

The bolt to the rifle was on the bottom of the receiver and when fired the empty shells would come out downward and hit the deflector ejecting the cartridges to the side. I have a habit of looking into the chamber to see if there were any cartridges. However, the shell deflector covered the receiver and made it impossible for me to see into the chamber. I pointed the rifle away from Greg, and after operating the bolt numerous times attempting to eject any unseen cartridges, I pulled the trigger. The gun went off, and the bullet went through the mirror on my dresser. Fortunately, the walls were concrete blocks which prevented the bullet from going through the wall and into the next room where everybody had gathered. My mother and father came running into my bedroom and was relieved to see us both unhurt, except for my pride.

Greg's MG was a TC model five years more senior than mine. It looked very similar to mine with the exception it had larger wire wheels, and the steering wheel was on the right-hand side opposite of mine on the left-hand side. Even though Greg's car was an MG, it was obviously different than mine. His spare tire was bolted to a bracket and attached to the exposed gas tank. The gas gauge was the same as mine; a green lite mounted on the dashboard that when lit, two gallons remained in the tank. His worked correctly, mine never did.

After our first trip to Hollywood, Greg invited me to accompany him to his Judo club which was a short distance from the University of Southern California (SC). Greg was a third-degree Brown Belt, and I watched as he practiced. The degrees of progression went from White Belt to Green Belt, Third Degree Brown Belt, Second Degree Brown Belt, and First Degree Brown Belt to First Degree Black Belt. The Black Belt has ten degrees. There was only one-tenth degree Black Belt in the world, and his name was Sensei (Japanese for Teacher) Mifune, living in Japan.

JUDO

Greg asked me to join his club called, Seinan Judo Dojo, meaning Southwest Judo Club. When I approached my parents and asked their permission to join Seinan, they told me no because they were afraid I would get hurt. If my parents had allowed me to join, they would be required to sign a release removing Seinan of any liability. I decided to join anyway, so I forge my mother's signature once again. Signing my mother's name was easy for me because I had done it many times while in school. When I did something wrong, the school would send letters home explaining what I did. She was supposed to sign acknowledging she read the letter, and I would take it back to school. However, I never showed them to her but instead signed them myself. After forging my mother's signature on the release, I started judo training as a White Belt.

After a few months of training, I had the opportunity to practice with a black belt named, Jean LaBelle. He was the United States Overall Judo Champion in Judo. I couldn't believe I had this opportunity and he had asked me to practice with him. He even allowed me to throw him twice, but most of the time I was flat on my back. I wanted to go home and brag to my parents but I couldn't because I hadn't gotten their permission to attend judo classes.

FIGURE 2 - MY FIRST VICTORY

After a few months of pestering them along with help from Greg's mother and father, they finally agreed to allow me to attend judo classes. They signed the release form, but I threw it away it because Seinan already had a release that I signed.

Learning judo was going to improve my ability to fight. After all, I was still a hoodlum at heart and learning to fight better was great, and I thought it was what I wanted. However, it didn't work out that way. I had never had my butt kicked in so many different ways and by so many people that eventually, my attitude changed and I was getting rid of the hoodlum ways I had developed in New York. I learned to love judo and how it changed me while growing up. It was the key element that healed me in mind and body. I still had the New York accent, and I tease everyone that I went to speech therapy to get rid of it. I lost it because everyone I hung out with either spoke Japanese or had a California accent. Attending Glendale High also helped.

Greg and I would go into Hollywood driving both our MGs him taking his TC and me in my TD. Greg's TC was a right-hand drive while mine was a left hand. We would drive in one lane of traffic with him on my left so that we would be alongside one another talking. We did this often, and we learned each other's driving habits. We didn't have to tell the other about what we were going to do next because we automatically expected the other to react. It was like flying formation in an airplane.

We were on our way home one night driving along Franklin Blvd which paralleled Hollywood Blvd to the north, and as usual, both of us were in the same lane talking. We didn't notice a police car following us about a block to our rear. As we approached Los Felix Road, we never signaled but turned left while alongside on another all the while both staying in one lane. The police car following was unprepared for this turn expecting us to go straight. The Police Car spun out of control while attempting to follow us into the turn. We heard tires squealing and realized what had happened. We both floored the accelerators and got out of the area as quickly as possible.

******

The procedure for conducting judo tournaments is that all participants are sorted by the rank of the belt worn and then by an individual's height. Everyone counts off by twos, (one two one two) and all the ones are one group and the twos the other group. Each group then faces one another, and one person from each group fights. The winning contestant remains on the mat and fights another contestant who is next in line. The process continues until one person is either beaten or ties.

One night when driving home from Judo, Greg, and another friend from high school, Greg Nibert and I drove into Griffith Park. We had stolen five gallons of concentrated liquid soap, and we planned to pour the liquid soap into the William Mulholland Memorial Fountain located at the busy intersection of Riverside Drive and Los Feliz Road.

Conspiracy

The base of this fountain was round approximately fifty feet across and had four slightly smaller tiers, each filled with water approximately two feet wide and about a foot deep. The base and main tier was the biggest and had a depth of two feet of water. On the top of the fountain was a large, round cement structure with water spraying upward for about twenty feet.

We poured the liquid soap into the water on the bottom tier and walked across Riverside Drive and sat on a park bench and waited. Naturally, we were nervous, so the time went by very slowly and for the longest time nothing happened. Then, once the liquid soap got into the more powerful fountains, it began to foam, and foam it did. When the soap reached the main fountain, it began to foam. Water squirting twenty feet in the air separated portions of the foam into balls which broke off from the main part and bounced along Los Feliz Road. Cars coming down the street slammed their breaks on not realizing what this phenomenon was. Some cars hit the foam balls, and they would splatter not damaging or injuring anyone. Eventually, foam engulfed the entire fountain and water went straight into the air out of the foam. It was impressively beautiful.

The Los Angeles Police finally showed, and when they realized the foam wasn't causing any damage so long as the cars didn't run into one another, they relaxed. They directed traffic around the area. When they saw the three of us sitting on a park bench watching the whole thing, they came over and talked to us. Naturally, we played dumb to the entire episode. Two cops had smiles on their face while talking to us and we think they knew we had something to do with what was going on. However, we continued playing dumb and eventually the police walked away. The foam eventually dissipated when the Los Angeles Fire Department with their hoses sprayed the entire fountain with water.

Since this episode, it has become an annual affair, and each year the fountain is again engulfed in a mound of soap suds. However, I do not know if this continues to this day.

After about two years practicing judo, I worked my way up to First Degree Brown Belt and had entered a judo tournament that held at a local high school. I had won my first match but tied with my second. I sat Japanese style with my legs under me and my feet crossed. When the tournament ended, I was unable to get up because I had strained one of my back muscles. I finally got up but couldn't straighten. For about a month I was stooped over and couldn't walk upright, but I didn't miss a practice session in judo regardless of the pain.

During this time frame, I entered a judo promotional tournament which held in the gym of a local high school. If I were successful in this tournament, I would receive my First Degree Black Belt.

My back was still very stiff and sore, but I vowed I was going to win my black belt. I figured I had enough strength in my back for one attack. I threw my opponent for the victory. However, I again strained my back, and I could barely walk. The tournament procedures were for me to stay standing and continue fighting until I was defeated or tied. My next opponent was a draw, but I scored enough points to be awarded my First Degree Black Belt. I was ecstatic over this promotion.

Members of the Seinan Judo Dojo were predominantly Japanese. They were very reserved toward Greg and me until they realized we were dedicated judokas. When they offered their friendship, it became obvious they were very sincere. When we attained our black belts, it became more obvious that we were part of the group. It was a great feeling and a great honor.

Santa Monica Beach was famous for their muscle building workout right on the beach. Women traveled from all over to come and watch the weightlifters bulge their muscles. The weightlifters enjoyed having the women watch them as they pumped iron. One weekend, Greg Knapp, Dan Powers, Rickey Tanaka, Victor Yoshimura, and I, all Black Belts went to muscle beach. We began practicing judo alongside these weightlifters. We hammed it up by throwing each other with one hand or sweeping a foot and have the opponent jump making it appear as if he was thrown. It was hilarious because it was obvious the weightlifters didn't like our company, but no one complained.

We used to hang out in a Bohemian restaurant where we could watch the weird customers come in. Then one day we realized people came into the restaurant to watch us because we dressed like no other. What a rude awakening.

While in high school, Greg and I attended a party with other students. Naturally, there was drinking going on. One friend was sitting in a chair with his girlfriend on his lap. They were kissing and making out heavy. I walked up and grabbed his girlfriend and pulled her off his lap and started dancing with her. After about a minute her boyfriend got up and tapped me on my shoulder and asked in a very stern voice, "May I cut in?"

I said, "Sure but I have to lead." I started dancing with him. He was so dumbfounded; he took about three dance steps, realized what he was doing and pushed me away.

When I first pulled his girlfriend from his lap, everyone thought there was going to be a fight. However, when I started dancing with the guy in my arms, everyone broke up laughing including this guy and the tension vanished.

******

My father leased the gas station for only three years. During this time local merchants became frequent and loyal customers. One merchant owned a manufacturing company that produced various wire products. I was contacted by the Pete the company's foreman; who set up a lease for my truck to deliver their products throughout the Los Angeles area. I would get paid by the location where I delivered these products.

Sometimes I would be gone from the gas station all day. The money I made was more than a normal sixteen-year-old kid would make, allowing me to live a high lifestyle. I sold my MG and bought a 1953 Jaguar Roadster, a car I had always wanted. It was white, had wire wheels and a red leather interior. The engine was a six cylinder with dual carburetors and dual pipes that created a very special loud humming sound. One day when entering the onramp for a freeway another car decided to race me. I went passed him doing over one hundred miles per hour when I shifted into high gear.

FIGURE 3-1953 JAGUAR 120 "C"

The drivers head turned immediately in my direction with a look of utter amazement and then immediately slowed down. I loved that Jaguar.

The Jaguar had a convertible top that I removed and stored it in the rafters of our garage. It gave me more room to store such items and my judo gear and dirty laundry. The trunk wasn't very large, but it did give me more flexibility for storing things. The car had a leather tonneau cover that would snap on over the cockpit area. When it rained hard, and without the top, I would pull over to the side of the road, pull the tonneau cover over me, snap it in place and wait until the rain stopped.

I enjoyed my cars. I had the Jaguar, 1951 Mercury stick shift with overdrive, a 1938 Ford four-door, and a 1941 Ford Coupe with a rake that raised the rear of the car. I also modified the stick shift by moving it to the left side of the steering wheel. It looked different but was not very practical.

While driving my brother's black Oldsmobile, three of us were returning home from Judo practice when I stopped at a traffic light where I was to make a left turn down a residential side street. This side street would take us past my father's gas station. The street we were on was six lanes, three going in our direction and three going the other direction. There was a police car in the center lane coming towards us waiting for the light to change. I told Greg, watch this and when the traffic light turned green, I made a left turn in front of all the oncoming traffic including the Police Car. My tires screamed as I burned rubber throughout the turn.

The reason I was driving my brother's car, I had loaned him my car so he could go on his honeymoon. Recently his car was severely damaged in the rear by someone who had rear-ended his car, so it wasn't hard to recognize.

We went down this two-lane side street at over one hundred miles per hour. The police car only had a six-cylinder engine, so it wasn't very fast, but he was following us. We pulled away from them very rapidly.

We turned right onto Chevy Chase Drive headed toward my dad's gas station and was approaching a railroad crossing. When we crossed the first set of tracks, we became airborne clearing the next set. We hit the other side rather hard but under control. We then crossed San Fernando Road, which is another main thoroughfare and the location of my father's gas station. The traffic light on the corner just turned red. Fortunately, we cleared all traffic without hitting another car. After two blocks we turned left onto a residential street and turned into a private driveway drove to just before the garage stopped and shut everything off. We thought no one would see our car. We left the car and started walking. Greg and I had worn our judo jackets which had bright orange sleeves and a black chest with a big orange emblem on the front with the letters JUDO TEAM spelled out in orange. How obvious can you get with our jackets?

We got two blocks from where we parked our car when the police showed up and arrested us. It was around eleven at night, and our parents had to come to the police station to pick us up. Why I ran from the police is beyond me, and even now I can't think a good reason for doing what I did. It was totally stupid.

When the owner of the house came home, the police thought it was us. They pulled their car behind the owner's and ran up to the car window with their guns drawn. Then they saw our damaged vehicle parked in the back and knew we were on foot, walking. Naturally, the people in the car almost had a coronary when the police came running up. It didn't take them long to find us wearing the orange and black judo jackets. They arrested and took us to the local police station, where our parents had to come to get us.

I had to go to court with my mother because I was under eighteen. The judge said he didn't understand why the police didn't shoot at us. I said I know why and when the judge asked why I said they couldn't get close enough. I thought my mom was going to have a heart attack.

The judge said he is going to have to suspend my driver's license and asked if I had any objections. I told him I worked at my father's gas station and I needed my driver's license to run errands for him and to make deliveries with my truck. He understood and said OK I could drive at the gas station. He then asked any other reasons? I said yes, I need my car to go to Judo training. He liked my working out at judo and agreed to allow me to drive to judo. It turned out I only had my license suspended on Sunday and Wednesdays when the gas station wasn't open. Also, I was allowed to drive four days a week to go to Judo practice. My mother couldn't understand my penalty, but I was happy. This suspension lasted a full month. The days I couldn't drive, Greg drove.

After three years of working at the gas station, my father decided to shut it down. He applied and was selected to become a distributor for Snap-On Tools. Greg went up north to attend Stanford University, and I joined the Marine Corps.

Chapter 3

U. S. MARINE CORPS

Glendale, California, on MEEay 16th, 1958 I enlisted in the United States Marine Corps. When I first enlisted the recruiter made a big deal about the date my birthday on November 10th because it was the same date as the Marine Corps Birthday, only the year was of course different. The year 1775 was when the Marine Corps was born, but mine was a lot later. They wrote an article in the Glendale newspaper pointing this out.

June 3rd I left Glendale for the Marine Corps Recruit Depot (MCRD-Boot Camp), in San Diego, California. The shock was astounding going from civilian life to boot camp where they strip you down by harassing, ridiculing, and demoralizing you until you don't recognize yourself any longer. Discipline was horrendous, and you couldn't say anything without addressing yourself as "private." "Private Adams request permission to speak SIR." Every sentence had to end with SIR, or else.

That first week in boot camp was extremely tense. As a kid in New York City, I had joined a military type of organization called "Aeronautical Cadets" and learned the fundamentals of marching which came in handy in the Marine Corps because the Drill Instructor (D.I.) recognized my marching ability and made me a Platoon Leader. There were four squads in a platoon. Each squad had four riflemen, three carrying an M1 Garand Rifles and one carrying a Browning Automatic Rifle (BAR).

During my first week, the drill instructor had us fill out some questionnaires regarding the things we did before joining the Marine Corps. I include my achievements in Judo as we had just completed a Judo Tournament before my joining the Marine Corps. Seinan judo club put up two teams in which every person held a black belt. I was co-captain of the second team, and we had to fight our first team for the California Championships. We lost to the first team but won a trophy for second place. Our Sensei, of course, was ecstatic because this was a first for him.

The entire platoon was ordered to fall out into a marching formation in front of the Drill Instructor's hut (Quonset Hut). When everyone was in place, the Lead Drill Instructor (D.I.) called me to the front of the platoon. He told me to about-face putting me facing the entire platoon.

He then announced that I was to be "Right Guide." The right guide was the leader of the entire platoon, second to the D.I. and when marching I would carry the platoon's red flag (Platoon 143- Guide Arm) on a pole in front of everyone. The Guide Arm weighed approximately four pounds, while the unloaded M1 rifle everyone carried weighed eight and a half pounds. The D.I. then announced if anyone wanted my position of right guide, they would have to fight me and win to take over my position. That first two weeks, I had numerous fights, but I held onto my position. If I could hold onto the right guide position until graduation, I would go from Private to Private First Class (one stripe). Even though I would fight a few of my fellow Marines, it cemented lifelong friendships and made my job easier.

Halfway through Boot Camp, we traveled from San Diego's MCRD to the Marine Corps rifle range in Camp Mathews just north of La Jolla, California. We spent two weeks practicing with the M1 Garand Rifle. At the end of our first week at Camp Matthews Rifle Range, we had open house allowing my parents and brother to come to Camp Matthews and visit. We had a picnic outside, and my folks brought beer in thermoses. Recruits weren't allowed to have alcoholic beverages, so we had to disguise the beer. While enjoying our get together and drinking beer discretely out of paper cups, one of my D.I.s came up to say hello and meet my parents. I think he smelled the beer but never said a word.

Twelve weeks of Boot Camp and graduation day was at the Marine Corps Recruit Depot. Three platoons of seventy-six recruits were graduating. My parents showed up for the ceremonies in the base auditorium. During the ceremony, there were three occasions when recruits went up on stage and received recognition for their achievements. The first award was for being honor man of their particular platoon. I won that award and received a certificate. The second award was for outstanding marksmanship with the M1 Rifle. I won that award also for shooting the second highest score in the platoon, and I received an expert rifle medal to wear on my uniform along with a trophy. The third award was for being promoted from Private to Private First Class. I won that award also and went up on stage to receive the citation.

My brother told me later that my parents were so proud they had tears coming down their cheeks. That made me happy because I knew they had doubts about me ever achieving any success in life when they compared to what I become growing up in New York City.

The entire platoon left MCRD by a Marine Corps bus, and we traveled to Camp Onofre, Camp Pendleton, at Oceanside, California. There we spent four weeks combat training. Upon arrival to Onofre, the drill instructor designated me as Platoon Sergeant giving me the same authority as I had in boot camp as Right Guide.

Part of our combat training required us to conduct maneuvers on a mountain called Old Smoky. We were conducting an exercise in which we used blank ammunition. It was during September and winter hadn't arrived yet so, temperatures were still high, and the grass was long and very dry. One Marine using a Browning Automatic Rifle (BAR) with blank ammunition did not consider the dry grass when he fired the BAR with flame shooting out the barrel causing a small fire which rapidly spread. A rabbit caught fire and ran through the grass spreading the flames throughout the area. Soon the entire hillside was burning profusely.

Our company commander announced we were not on fire standby, so we left the mountain to burn on its own. Fortunately, no one was injured, and the company that was on fire standby had to climb the mountain and fight the fire. The fire had burned for three days before they were able to get it under control.

After combat training I returned to MCRD to wait for the Communication Technician School, I had been selected to attend began in Imperial Beach, California, located about nine miles south of San Diego. This specialized training required me to have a Top Secret Cryptographic Clearance because the subjects learned had been classified by Company G (Intelligence) in Washington, Headquarters. It took the FBI about six months to complete my background check.

Training lasted five months, and when completed, our instructors listed the locations of duty stations we could serve at and we selected the one we wanted. I chose Japan so that I could continue my Judo training at the Kodokan, headquarters located in Tokyo. The Kodokan is where most of the champions of the world trained except Mr. Geesink. He was the only Caucasian to hold the world title two times. He was Dutch.

I graduated number two in my class of Technicians and sent to Karamursel, Turkey, the opposite side of the world from Tokyo. I should have requested Karamursel, Turkey and they probably would have sent me to Japan.

TRAVELING OVERSEAS

I learned we were going to travel to Turkey by ship. I traveled from California to New York City in a Douglas DC-7 commercial airliner and reported into the Brooklyn Naval Ship Yards. That weekend my Cousin Tommy and I boarded a train heading to Cape Cod, Massachusetts where he owned a vacation cabin and his family was there waiting. The train left us in Boston about seventy miles from Cape Code. From there we hitchhiked to the Cape.

We were picked up by a Marine Corps Captain who was also heading to Cape Cod. I was very concerned because I was beyond the limit of a weekend pass. So, in essence, I was absent without leave (AWOL), and if caught I would be locked up in the brig. Fortunately, the Captain never noticed I had Marine Corps Uniform Shoes.

We stayed in New York for a week, and I showed everyone around the city where I grew up. We had a great time, but I don't think New York City did.

We boarded a troop ship destined for Istanbul, Turkey. The first day out on the ocean, the ship was rocking back and forth, and people were not used to the rough water. We were standing in very long line at the entrance to the mess hall (cafeteria) waiting to eat when an individual rushed up to the trash can and vomited. The long line we were in immediately became a short line.

We crossed the Atlantic Ocean in six days and felt fortunate we didn't run into any bad weather. Our first stop before going through the Straights of Gibraltar into the Mediterranean Sea was Cádiz, Spain. We weren't able to leave the ship because the captain was afraid we wouldn't get back on board in time to sail. Our stay was short and disappointing.

The following day we stopped at Barcelona, Spain where we did leave the ship for the first time. We were, however, expected to be back on board before nightfall. From there we went to Livorno, Italy or otherwise known as Leghorn, Italy. We were only 31.6 kilometers (20 miles) from the Leaning Tower of Pisa. Five of us decided to make the trip to Pisa by bus. It was our first experience in Europe, and we didn't know what to expect. When we boarded the bus, we moved to the side so that a gorgeous woman could board. When she reached her arm to take hold of the handgrip, gobs of long black hair was seen coming from her armpits. We looked at each other and somehow kept from laughing but, she was beautiful. The Leaning Tower of Pisa was interesting. How it remained upright, I'll never know.

We stopped at a small café to get something to drink and ordered a Pizza. The Pizza was the worst I have ever tasted. We didn't even finish it. It was thick and doughy and not very tasty. One of our group, Tom Wells started talking to a girl who was at the next table. She spoke pretty good English and was very friendly. Tom was half drunk and kept hanging onto her like she was going to run away. I took a picture of the two of them together. I wanted to show him after sobering, how ugly she was. I have since destroyed the photo because she was a nice person.

FIGURE 4 - REMNANTS OF WWII LIVORNO, ITALY

FIGURE 5 - CASTLE NUOVO – NAPLES, ITALY

When we got back to the ship, two Italian dock workers were loudly arguing, using their hands and fingers for emphasis. We laughed listening to them argue.

Leghorn still had buildings that were half bombed out from WWII. Dilapidated factories with wrought iron window frames on the roof bowed upward from where a bomb had exploded inside. The walls around the factory had rusty barbed wire on the top and bullet pockmarks were everywhere. The only improvement was the debris had been removed, and only the structure was left. It was sad to see the destruction.

We next docked in Naples, Italy. From the ship, we could see Mount Vesuvius in the distance. We left the ship and traveled to the military base on the outskirts of the city. We had lunch in the cafeteria, and I ate real Italian Lasagna. The food was outstanding. We stayed most of the day while the ship took on provisions.

The next port we arrive at was Tripoli, Libya. Tripoli is in the Marine Corps Hymn, and it was exciting being there. The Marine Corp was born when President Thomas Jefferson sent our warships with Marine Rifleman on board to combat the Somalian Pirates who were raiding ships along the northern coast of Africa.

All the buildings we saw were a sand color which was the same color as the streets. The contrast between all the tan buildings and sand, and the deep blue color of the sky and the Mediterranean was quite beautiful.

The next day we were in Athens, Greece. We were in port for three days and took advantage of seeing the sights. Five of us traveled to the Acropolis of Athens and saw the Parthenon, Erechteion, Temple of Athena Nike, Plaka, and the Acropolis Museum.

We left Athens and sailed to Istanbul, Turkey. We arrived in the Bosphorus just outside Istanbul, late at night. Just behind and very close to the stern of our ship another ship turned on their bright lights and started blowing their collision warning horn. That ship swerved to the right and passed. When they got a beam of us, we saw their Russian Flag. I don't know whether they were playing chicken or we had an actual close call by them ramming our rear.

ISTANBUL, TURKEY

By the time we reached Turkey, we had visited in every country bordering the Mediterranean Sea.

We took a ferry across the Bosporus, climbed onto a military 6X6 truck, and headed east, around the Sea of Marmara, to Karamursel, Turkey, and the Air Base.

The Air Base had Navy, Army, Marine Corps, and Air Force Personnel. Except for the Marine Corps, all personnel had their barracks with private rooms that held four people. Some of the personnel had rooms with only two people and single bunks, while the rooms with four people had two bunk beds, one person slept on top and the other on the bottom.

The Marines slept in eight Quonset huts which located at the farthest point west of the base. We had single bunks and didn't have to sleep above or beneath anyone. Everyone on base named the group of Quonset Huts the "Quiet Village" in remembrance of the famous song. When we had parties, people at the main base heard weird noises coming from our Quonset Huts, so they called our area "The Quiet Village."

The base Provost Marshall was an Air Force captain who also held a Black Belt in Judo. When the base held Smokers, (amateur boxing matches) he and I would put on a Judo Demonstration during intermission. We had a routine we rehearsed, and everyone like it. We did this quite often and established quite a reputation. However, it did have some disadvantages.

FIGURE - 6 MARINE CORPS DETACHMENT

Conspiracy

When drinking at the Enlisted Men's Club, I would be sitting with my friends when someone unknown to me would come up from behind and hit me. This someone would get bigger and meaner with each drink until he just had to test me. The more they drank, the tougher they became, and they looked around to see who they could punch, and they would see me. I wouldn't be bothering anyone when all of a sudden I would get punched in the back of the head, and it would start.

During one of these tussles, my friend Mitchel went over to the line of slot machines and picked up a quarter slot and was headed for the exit when two Air Police came walking in. Mitchell stopped in his tracks, made an about-face and replace the slot machine on the shelf. The scuffle immediately stopped. No one got arrested.

The base had a beach on the Sea of Marmara. Families could go there and lie around and get a tan. However, if I did that with my fair skin, I would burn, peel and then get burned again.

There was also a small fourteen-foot runabout boat with a Johnson Outboard Motor on the back. It was big enough and powerful enough to pull two water skiers.

On my first day of learning to waterski, I was doing pretty well when something caught my eye off to my right. I looked and saw two fins break the water about twenty feet from me. I almost fainted thinking they were sharks. There was no way I was going to lose my balance and fall, so I didn't let go of the tow line and hung on for dear life. When I had a chance to turn my head again, I saw those two fins were not sharks but belonged to two porpoises. I felt very weak but decided I was going to survive after all.

We had a beach party where numerous Air Force, Navy and Marine Corps personnel were in gathering. We had built a huge fire out of old boards and were making a lot of noise. The Air Police showed up demanding we break up the party because people were complaining. We agreed and decided to move the party to the "Quiet Village" (Marine Corps Quonset Huts. We started breaking up the fire by pulling the burning boards off one at a time. I pulled two boards nailed together, and when I studied the structure, it looked like a cross. The boards were burning at the end that had two boards in the shape of a cross. The lower portion left over three-quarters unburned and clear to hold.

I thought it was really neat so I picked up the boards and raised it into the air. We all got the inference and started walking holding the burning cross. We then started chanting as we walked. The Marine Quonset Huts were on the other side of the base about half a mile away. It was dark and when we walked past the movie theater; three black airmen were just leaving and saw us. When they saw the burning cross and heard us chanting, they turned and ran like hell. They didn't even notice one of the people helping me carry this cross was black. We all laughed like crazy.

The base commander got upset because he thought the entire episode was racially motivated. He didn't realize several black airmen were walking with the group holding the cross. He never found out I was the ringleader.

FIGURE 7 - DOMABACHIE PALACE

I started going to Istanbul every time I got a couple of days off. We would take a ferry across the Sea of Marmara and then ride a bus to the east side of the Bosphorus. Then we would board another ferry to the other side where two of us rented a small apartment just off Taksim Square.

Conspiracy

FIGURE 8 - THE BLUE MOSQUE AND SAINT SOPHIA MOSQUE

Taksim Square was an intersection where five roads came together and met in a roundabout. In the center of this roundabout was a statue of Mustafa Kemal Ataturk who was a very famous Turkish General who fought in the early 1900s and hero of the Turkish people.

All the Americans frequented a bar and dance hall establishment called the Caravan. The Caravan, owned by Mustafa, was one of the leaders of the Turkish Mafia and who has no bearing on anyone today. Mustafa heard of my judo experience and asked if I would wrestle someone he knew. I said I would, and he set up the match. My opponent was the Turkish champion in his weight class which was just about what I weighed. His physic was that of a Greek God, small waist, and massive shoulders. We fought, and I won. Mustafa was totally surprised and congratulated me over and over. During the match, I had ripped my opponent's shirt to shreds, so I bought him a new one. Mustafa thought this was a wonderful gesture. That was the last time I saw my opponent.

Every time I went into the Caravan, which was often, Mustafa would come over to my table and talk. One night he brought his brother over and introduced him to me. I later found out Mustafa's brother was the hit man for Mustafa's organization and that he had killed eight people. My problem was, he took a liking to me, and every time he saw me he would come over and start talking. Fortunately, he didn't speak English, and my Turkish was limited, so these conversations didn't last long. I'm normally not afraid of anyone. However, with this guy it was different. He was short and stocky but not fat. He was completely bald and had steel blue eyes, which is unusual for a Turk. When he looked at you, it appeared he looked straight through you and focused somewhere behind me. It made me nervous. Even though he liked me, I would rather have been around someone else.

Istanbul means "Lonely Planet," and its history is unique. Built on six mountains, the people have very muscular legs from walking up and down the streets. We visited the Saint Sophia Mosque just south of Istanbul. It is the oldest Mosque in existence and built between 532 and 537 AD. It was a Christian Church at the time and is the reason it has maintained its title Saint Sophia. During 1453, under the rule of Sultan Mehmed II, he ordered it converted to a Mosque. It had four spirals built in its conversion from Orthodox Christianity into a Mosque. In 1935 the Mosque was turned into a museum and open to the public. The Masque structure is of solid marble. The entrance floors are concave from so many people entering and leaving.

Saint Sophie Mosque was the main Mosque in Istanbul until 1616 when the Blue Mosque came into being. The Blue Masque is the only Mosque in the world that has six spirals. We called these spirals, God Rockets.

The city has a very interesting shopping area called the Grand Bazaar. In Turkish it is called Kapalicarsi meaning 'Covered Bazaar'; and is one of the largest and oldest covered markets in the world, with sixty-one covered streets and over 3,000 shops.

In 1961, while I was on leave in Istanbul, the military overthrew the government. We couldn't get back to base for four days. Tanks were going up and down streets, and three tanks blocked Taxim Square two blocks from my apartment. The Revolutionaries hung a copy from the Golota Koprusu (Koprusu is Turkish for Bridge). The military was everywhere and carried a machine gun, and you could see the bolt in the fully automatic position with cartridges showing. Many people were arrested or shot. Curfew was at 8:00 PM, and we made sure we were not on the streets when the curfew was in effect. After four days of hell, things returned to normal, and we went back to base.

TRAVELING HOME VIA AIRPLANE

After a year and a half, my tour of duty in Turkey was over, and I returned home from Istanbul by airplane, again a DC-7. Our first stop was in Cyprus. In route we encountered such severe turbulence, the airplane captain admitted it was the most turbulence he ever encountered.

My seat belt was fastened loosely around me, and when we hit the first bump, I flew up, and the only thing that prevented my head from going through the roof was the loose seatbelt. I tightened it up as tight as I could get it. Looking out the window, I could see lightning all around the airplane, and it scared the hell out of us.

We stayed in Cyprus long enough to eat and refuel the airplane. Departure from Cyprus was uneventful because those thunderstorms we encountered fortunately left the area. The flight between Cyprus and Porte Lyautey, Morocco was uneventful. We stayed two weeks doing nothing. I guess the powers felt we served enough, in Turkey, so all we did was go into town and drink Moroccan Bear.

Our next stop was the João Paulo II Airport on the Azore Islands named after Pope John Paul, II. We stayed just long enough to fuel and get something to eat. I heard my name being announced over the loudspeaker summoning me to the military station. When I arrived, I was asked if I had a Top Secret Clearance which I did and I was given a .38 revolver in a shoulder holster and asked to accompany my companion a Lt. J.G. to a security door. We received a heavy package and were asked to deliver it to a courier who will meet us when we arrived at New York International Airport. (Future name to be J.F. Kennedy Airport)

After becoming airborne, an attendant came up to us and asked if our guns had bullets in them. I answered affirmatively. The attendant then asked for us to unload our weapons and give the ammo to him. We both told him to git because he didn't "need to know" about the classified data we were carrying. He very sheepishly walked away, and we weren't bothered again for the rest of the flight.

Our flight number was the same as one assigned to a flight that departed the Azores a month prior ours. That airplane vanished from radar approximately three hundred miles east of Newfoundland. On board were three buddies stationed with me in Karamursel, Turkey. They never did find any trace of that airplane. All across the Atlantic, we all had a weird feeling until we landed in Newfoundland, just off the coast of Novia Scotia, Canada.

Our next stop was in New York City at the then New York International Airport. Once we deplaned the Lt. and I met a carrier and got rid of the package we were carrying and said our goodbyes and we went our separate ways.

COMING HOME

I boarded my flight to Los Angeles, California and I spent the next thirty days at home in Los Angeles before beginning my last year in the Marine Corps.

While I was in Turkey, my parents bought a bar on Sunset Boulevard just south of Dodger Stadium. My mother would work from eight o'clock in the morning until five pm. Then my father would take over and work until two am. They only sold beer, but it was quite popular because they had a special glass cooler. It had frozen cones in the shape of the glasses they used which would, in about three minutes, freeze the glass. Filling the frozen glass with beer would make the beer much colder. Construction workers would come into the bar and drink from the frozen glasses. The hot summer months brought many workers into the bar, and they would drink beer for hours.

Coming home from a judo practice, I would go immediately to my folk's bar and drink three glasses of cold beer before heading home. It helped me sleep as if I needed any help.

In October of 1961 and on occasion I would help out by tending bar allowing my folks to get away together. One night, while tending bar, Ron, one of the customers asked if I was going to the Halloween Party. I told him yes and that I would see him there.

When I arrived at the bar on Halloween, I could barely get inside it was so crowded. Once inside, a cute gal came up to me and said hi Al. Dressed in a tight sweater and skirt, and with her tantalizing cleavage showing, I mentally assumed my night was going to be complete when it became obvious to my newly acquainted friend, that I did not recognize her. She asked me if I recognized her, and when I look puzzled, she told me her name is Ron. I then recognized who I was talking to and it was at that moment I realized everyone or just about everyone in the bar was gay. I was naïve to the gay crowd and their ways, and this took me by surprise.

I still tended bar to give my parents a break and became good friends with everyone. I was even invited to accompany four of them to visit some hangouts in Hollywood. While in Hollywood, and after about thirty minutes in one bar just off Hollywood Boulevard, I noticed I was alone sitting at the bar with no one sitting next to me? My friends had formed a semi-circle around me keeping all strangers away. I told my folks about it, and we all laughed, but we were proud of my new friends. The evening was fun and entertaining.

MARINE CORPS JUDO TEAM

FIGURE 9 - MARINE CORPS JUDO TEAM

At the end of my leave, I reported to my new duty station at Camp Lejeune, North Caroline. Upon arrival, I would be temporarily assigned to Quantico Virginia to participate in the All-Marine Judo Championships. Quantico is where the Marine Corps had the Officers Candidate School and the FBI's Training Academy. My company commander, Captain Burin said I wasn't going anywhere because I had just gotten there. I told the judo team about my predicament, and Captain Burin received a call from the Camp Lejeune Commanding General's office. The next thing I knew, I was on a bus heading to Quantico, Virginia. I had never before met any of the judokas, and it was strange they allowed me to join them because I could have been a fraud. It seems a past judo student I taught in Karamursel, Turkey told the Lejeune judo team about me and they arraigned to have me join them in Quantico. What surprised me was they didn't even know if I was a black belt or someone who just claimed to hold that rank. Sometimes this happens, and it can be dangerous because when someone new appeared at my home Judo Dojo, Seinen, I would ask them to work out and if they weren't what they claimed, they would get their butt kick really bad. However, this didn't happen too often though.

I met Ernie Cates under these same conditions and who just returned home from Okinawa. He was a friend of Dan Powers who was the existing US Marine Corps overall judo champion. I challenged Ernie to practice, and within a few minutes, I had thrown him winning the point.

The judo tournament was a two-day affair. To win a judo match, you have to throw your opponent; to land flat on the back getting you one point. If your opponent landed on a side, it graded as half a point. You could also win a match by holding your opponent down on the mat for thirty seconds or have them submit by pressure from an armbar, or by choking. The choke hold is only authorized if you use your opponent's Judogi to apply the chokehold and not use your hands around the throat.

The captain of the judo team asked that I lose five pounds so that my weight would qualify me for being less than one hundred and seventy pounds. I starved myself and ran everywhere to lose the five pounds. The strategy behind this was to put me in the same weight class as Ernie Cates whom I had a thrown for a victory before I entered the Marine Corps. However, I had not practiced judo to any great extent for over two years, and I was just getting back into it when this tournament came up. I felt inadequate.

The first thing we did when we arrived was to put on our judo clothes. Each of the judo team worked out with me to test if I was what I claimed. No one could throw me, and I threw their captain, Sargent Bonar. He was good and tried to wrestle me down to the mat for an armbar. He did get me in an armbar, but I twisted out of it but not before I felt a strain in my elbow.

The day of the tournament I went undefeated, winning all of my matches. We had to weigh in again the next day, and I starved myself and ran for about an hour. When I weighed in, I had lost another two pounds, and I weighed in at 167 pounds. My weight was the lowest I had weighed since high school.

Conspiracy

The second day I had won all my matches until I met up with Ernie Cates. He was unable to throw me, but I lost when he held me down on the mat for thirty seconds giving him the championship.

At the end of the tournament, each of the champions in their respective weight class was selected to represent the Marine Corps in the United States Championships held in San Jose, California. I was to be the fifth member of the team because of my outstanding performance throughout these matches beating the other second place contenders.

I arrived back in Camp Lejeune and informed my commanding officer I would be going to San Jose, California for the United States Championships. This time he didn't object. Later that week, we all boarded a DC-3 out of Cherry Point, North Carolina. Our destination was El Toro Marine Corps Air Station, California.

Approximately halfway across the country, we landed in Kansas City, Missouri for fuel, and lunch. I couldn't believe how slow this flight was. Another five hours and we finally landed in El Toro. We spent two days at El Toro before heading to San Jose. My folks picked me up, and we had dinner and then home. It was only about two weeks since I was last home after returning from Karasmursel, Turkey.

Arriving in San Jose, California we all checked into our respective hotel and made plans for attending the championships the following day. We had been issued a Judogi with the Marine Corps emblem on the front and had our pictures taken. Then we began practicing.

The championships matches were two days in length. There is no telling how many matches you fought on a given day. It all depended on the luck of the draw. I went undefeated on day one, and two of our team members lost their matches. The second day I went undefeated again until I lost the match to an old friend Harry Fakuwa. I hadn't seen Harry since joining the Marine Corps. All my opponents I scored points on during the second day of the nationals were higher in ranking than me either second or third-degree black belts; I was only a first-degree black belt, so I didn't feel too bad. Having progressed into the Semi-Finals, positioned me as eighth in the United States standings in my weight class.

When we got back to Camp Lejeune, the local newspaper wrote a story about me going further than anyone else in the Marine Corps and that I placed eighth in the United States in my weight class.

About a month later, Camp Lejeune had a promotional judo tournament in which I was forced to participate. I tried to pass this tournament up but the team I went to California with forced me to enter. I found out their reason was that I had beat so many judo people of higher ranking they had to promote me.

They lined up eight judokas into what is called a slaughter line. The Slaughter Line is a type of match where one individual (me) faces six to seven judo opponents. I fought the first person, and if I win, he /she sat down, and I faced the next opponent. I continued fighting until I beat all my opponents or I either got beat or tied. I won all my matches and was awarded outstanding judoka of the promotional and received an impressive trophy. To keep the trophy from being broken, I placed it on the cabinets in the head office where I normally worked. Later, I mailed the trophy back to California where my folks put in on the shelf behind the bar for everyone to see.

I volunteered to go overseas again aboard a Landing Ship Tank (LST). I was in charge of a six by six truck which had a fourteen feet by seven feet shack on the back filled with radio and teletype equipment. Inside the shack were three radios permitting us to communicate with just about anyone within two thousand miles in Morse code, Teletype, or Voice transmissions. A gasoline power unit, weighing around seven hundred pounds was attached to the rear of the truck.

The ship had a flat bottom with no keel which allowed the ship to put its nose on the beach. Two huge doors made up the ships bow and when opened exposed a ramp going up to the main deck. I engaged the four-wheel drive and drove the truck into the nose of the ship and up a ramp to the main deck where we tied it down with chains to the main deck with three-quarter inch turnbuckles. It was impossible to move when the ship was on the ocean; this was mandatory to prevent it from crashing into other vehicles or being thrown overboard.

We left port and sailed into the Caribbean Sea heading for a small island called Viegas just off the coast of Puerto Rico. The ship had numerous Marine Personnel on board, and every morning they came on deck to do calisthenics. There were four of us; Corporal Robert Ott was our leader, Kevin as a radio operator and Charlie Bear as technician and myself. Each morning, after breakfast, we would go to our truck, open the door exposing the radios and equipment to anyone who happened to look in. One morning a First Lieutenant came to the door and asked why we weren't doing calisthenics with the rest of the Marines.

If you hadn't had any experience operating the equipment inside this van when you looked in and saw the radio equipment along with the teletype machines you would be confused. We told him we were guarding cryptographic equipment and if he didn't have a Top Secret Clearance along with a need to know, he had better leave the area. He immediately turned and walked away. We were never bothered again. Of course, it was all a lie. We just didn't want to do calisthenics.

Arriving at the Island of Isla De Vieques, Puerto Rico, and the LST got as close to the beach as possible but, the ship was still about one hundred yards out in the water. Metal floating ramps were unloaded and placed between the bow (front) of the ship and the beach. They were then tied together to keep them from drifting apart. I drove the truck down the ramp, onto the floating ramp and then on the beach. The truck didn't get one hundred feet onto the beach when all ten wheels sunk into the sand, and we became stuck once again. A Caterpillar Tractor had to come with a tow chain to pull us free.

We positioned the truck on top of a hill for better transmitting and receiving capabilities. Early the following morning, a Major came and asked if we could get communication established between us and headquarters located in Florida.

Our encampment was a mess with dirty laundry all over the radio gear. The major did not seem upset because of our dirty laundry. When the Major explained what was needed, we went into action.

It seemed that all their communications had failed. We loaded the necessary frequencies into the radios, raised the antenna, and immediately established communication when everyone else failed. I handed the major the message I had received and left. After a while, e came back and congratulated us for a job well done. We felt pretty good.

Based on the communications we received we left the island and went to Guantanamo Bay, Cuba. There we left the truck on the ship, went ashore, and had lunch in the mess hall. We then got word to head back to the ship because we had been ordered to stand guard off the shore of Venezuela to protect President Kennedy while he visited Caracas.

After leaving Venezuela, we sailed to San Juan, Puerto Rico and spent the next ten days celebrating Christmas. The four of us visited Morro Castle which was the fort protecting the harbor. It is now a museum. I remember walking down this tunnel passage and seeing a huge piece of shrapnel sticking out of the roof. One building had numerous old cap, and ball rifles in racks and they looked as if they were new. We located and visited a Judo Dojo, and I practiced with the local judokas. One of the Sensei's owned a bar in downtown San Juan where we went after practice. He appreciated us so much for coming and working out, he wouldn't allow us to buy a drink all night, and we drank many exotic rum drinks.

What a gentleman. We did this on one more occasion. We were walking down a dark street looking for another bar to visit when I spotted a Ford Convertible sitting in the middle of the street with the engine running. I jumped into the driver's seat and drove off. When the others caught up, I was around the corner waiting, but they wouldn't get in, so I abandoned it and walked away.

In March of 1962, we returned to Camp Lejeune and the ship we were on docked on the beach to unload. As I drove the truck down the ship's ramp, a Marine Corps Captain jumped on the running board of my truck and asked for a ride. I granted his wish but told him we didn't have any brakes. I lost the brakes putting the truck on the ship when we were leaving Vieques. He looked nervous and grabbed the side of the truck much harder as we descended the steep ramp. However, we arrived without incident on the beach where we once again got all wheels of the truck and trailer stuck in the sand and had to be pulled out by a Caterpillar tractor. The Captain wished us good luck as he jumped off the running board looking much relieved. Two weeks later I received my discharge, and I headed to California and my folk's bar.

Chapter 4

RETURNING HOME

After arriving home from the Marine Corps, I took about a month off and just worked out in judo about four times a week. After judo, I would visit my folk's bar and have a few beers. I went home a few nights feeling no pain.

I submitted applications to the Los Angeles Police Department and the Los Angeles Sheriff's Department. I waited and heard neither was hiring due to budget constrictions. I then applied to the AT&T Bell Telephone Company. I passed all the tests with high grades and got hired and assigned as a telephone installer in Beverly Hills, California.

I have always enjoyed electronics, and it was an interesting job in which I advanced rather rapidly from a single line telephone installer to a PBX installer and a PBX Repairman all within a year and a half.

Being assigned to Beverly Hills, we worked on many of the movie actors homes. One assignment was at Frank Sinatra's residence where we installed the closed circuit TV in his theater for the Paterson vs. Liston Fight. We were there two days, and each day Sinatra had someone bring us food and soft drinks. He came home on the last day and talked to us. He was nice to all of us and gave us each $20 tip for the job we were doing.

The best time I had was meeting the celebrity, Charlton Heston. He ordered another telephone number, and we had to install a larger cable from the telephone pole to his house. He came outside and talked to us while we put up the new cable. Soon he was helping us put up the cable and being in the middle of the summer we started sweating profusely including Mr.

Heston. He worked like crazy, and my helper and I appreciated his assistance. His garage was open, and I could see his Jaguar E Roadster. I still owned my Jaguar, and we had a great talk about the two cars. I had raced mine right after I first got it but quit when I almost crashed.

It wasn't until later I heard he had a fabulous gun collection. I also had a gun collection that I thought was second to none but limited to the total number of guns I could afford. I did have some rare ones that I was proud of, like a pair of Nickel Colt Single Action .45s with consecutive serial numbers, 10091 and 10092 and a rare DWM German Luger with an American eagle on top of the receiver.

Working for the telephone company I met my first wife, Jeanne Urricariet. She was full blooded Basque, and her grandfather was from the Pyrenees Mountains of Spain, in the Basque country. The town her grandfather came from was called Latasa and which was his last name. He was very wealthy and owned half the land along with Howard Hughes before it became Marina Del Ray.

Jeanne was a sales representative for AT&T in the business office. On numerous occasions, I had to call in to provide her changes to a customer's installation bill because of modifications I made in the installation and which had to be approved by her. We struck up what I thought was a lasting relationship. We always spoke on the telephone but never had met one another until she came to a party at my apartment on Saturday evening. Within six months we were married.

I stayed with the telephone company for two years before the Los Angeles Police Department finally called me to continue my employment process. Then, the Los Angeles Sheriff's Department also called to continue the employment process. I passed all the tests with the Police Department. They turned me down because I had a defect in my hearing along with a defect in my back. According to the doctor who gave me my physical, my vertebras were supposed to look like the outline of a Scotty dog. This disease would eventually present a problem with my back. They called it some disease, but it disqualified me.

The Sheriff's Department hired me, and I went to their Academy at Biscaluz Center. During my training at the Academy, I would ride in a radio car patrolling various parts of Los Angeles County. My first assignment was at the San Dimas Sheriff's Station. My partner was a veteran that had been in the Sheriff's department for quite some time. When we got in the radio car to go on patrol, he grabbed the steering wheel and yelled, "You're all animals and were out to get you." I thought he was kidding, but I found out later that night, he wasn't.

The academy class required every deputy to qualify with a revolver. Semi-automatic pistols were against the department policy. My preferred weapon was the Colt .357 Python revolver. It had the smoothest action of all the manufactured pistols. To qualify with the pistol, an individual had to shoot once a month and had to get a minimum score. For Distinguish Expert you had to achieve a score of between 290 and 300 points on three separate occasions. 300 was a perfect score, and expert was from 270 to 290 points, and sharpshooter 250 to 270 points and so on. On three occasions I shot 298 out of a possible 300 points which qualified me as Distinguished Expert.

FIGURE 10 - DISTINGUISHED EXPERT BADGE

Shooting has always been natural for me. I started shooting .22s in competition when I was twelve years old in New York City. I won numerous awards, medals, and trophies. After I had moved to Glendale, California, I bought a pair of Colt Single Action .45s with consecutive serial numbers and practiced quick draw and shooting from the hip. I learned to reload my ammunition which saved me a lot of money because my brother and I would go out to the desert and practice constantly. I got so proficient I could take a can or clump of wood and throw it into the air, draw my pistol and hit it before it hit the ground. My best score practicing this was four hits out of six shots.

The academy training was twelve weeks long, and upon graduation, I got assigned to the County Jail in downtown Los Angeles. The building was old and very depressing. It had eight floors with the courtrooms located on two of the floors, the men's cells on two floors and the women's jail on one floor. That was the most depressing job I have ever had. It was terrible the way prisoners had to live behind bars. There weren't enough beds for all the prisoners and so some had to sleep underneath the bunks and on the floor inside the cage surrounding the cells.

On one of my shifts, I was working in the booking room on the ground floor taking fingerprints of prisoners as they came in after being arrested. All of a sudden I smelled something so vile I almost gagged. In walked this man so dirty and un-kept the crotch of his pants hung down to his knees. We called these individuals "stinkos." They lived on the streets of Los Angeles and didn't eat any food. They drank wine until they got so saturated with alcohol, they couldn't even relieve themselves properly, so they did it in their pants. When people wouldn't give them money for wine any longer, they would get arrested, for some small crime and put in jail. Society cleans them up by providing showers, fresh clothes, and meals. When they serve their time, they go back into society, and the cycle starts all over again.

We had trouble with three prisoners just before this "stinko" arrived to be booked into the jail, and we had to use a little force to get them to cooperate. One of the prisoners punched a deputy. The "stinko" was being fingerprinted and processed while I called up to the jailer on the sixth floor and told them not to send the elevator down. When the deputy was finished fingerprinting the "stinko." we put him in with the three troublemakers. Soon afterward, the three wise guys started yelling to get this guy ("stinko") out of here. We did not move and let them stay there for a while. The yelling had continued for some time before we called for the elevator to come down and get them.

Fortunately, I only stayed in the jail for about three months before being transferred to the Sheriff's Station in Montrose. Normally a new deputy had to stay in the jail for a year before getting into a radio car. I had replaced a deputy that had answered a family dispute and was shot by a shotgun blast through the front door of the house. He later died in the hospital.

Conspiracy

The Montrose Sheriff's Station had two radio cars. Montrose 121 and 122. On weekends and special occasions, a Montrose 126 was activated to handle any special assignments. On weekends this car patrolled the mountainous area along Angeles Crest Highway up to Wrightwood.

During the winter months, the ski areas were open, and Montrose126 was a busy car. If the sun was out and no wind, skiers would wear short-sleeved shirts, and on occasion, women would be in bathing suits. However, once that sun went behind the mountains, the temperature dropped rapidly.

During a three to eleven evening shift, we received a call from dispatch, "Man with a gun" along with the address of the location. We contacted the other car on the tactical frequency (car to car) and told them to approach from the east, and we would approach from the west. We then kept in radio contact so that we both arrived at the same time. When we arrived at the location, the other car was coming towards us. Both cars stopped, and I grabbed the shotgun from its holder and opened the door. The other deputy in the other car was climbing out the passenger door with the shotgun pointing toward the sky. He loaded a shell into the chamber, and it went off (fired). It made me jump, but I noticed a severed telephone cable come crashing to the ground. Then we heard a voice coming from the rear yard of the house given to us on the radio call, "Don't shoot, I give up." I'm surprised we were able to subdue the suspect for we were laughing so hard and the shooter (deputy) was all embarrassed. From then on, we called him "Boom Boom."

It was right around the time of the "Union Fields" where two Los Angeles Police Officers were kidnapped and taken out into a farm's planted fields, and one died, and the other survived to identify the killers. When we heard this, many of the deputies started carrying an additional gun. These guns were small, and many carried derringers. The calibers were also pretty small and didn't have much power, but mentally, the officers feel more secure.

These guns, however, caused some accidents. We used to stick the radio car keys in the large buckle of our gun belt (Sam Brown). One deputy visiting the downtown headquarters building pulled his car keys from his gun belt, and his hideaway gun came out along with his keys, and it went off when it hit the floor putting a bullet into the ceiling. Fortunately, no one was injured.

The Montrose Station had a small booking room with two six-foot tables along two of the walls. I was sitting on one table when two Deputies brought a prisoner in to be fingerprinted. This suspect was drunk and very belligerent and kept yelling he wasn't going to cooperate. I noticed the watch sergeant taking off his Sam Brown gun belt, and I just knew blood was going to go all over. As the suspect passed in front of me, I reached over and grabbed the prisoner in a wrist lock and forced him to go to the floor face down. I held him there and made him scream in pain. I asked if this hurt and put a little more pressure on him. He screamed and yelled, "Yes, yes." I did this all the while still sitting on the table. I started thinking; I didn't want anyone to think I was showing off, so I slid off the table without letting go of the prisoner. Whispering into the prisoner's ear, I told him the pain he felt was nothing compared to what I could inflict, so behave yourself. He became very docile and submitted to the taking of his fingerprints. Everyone knew I had a black belt in judo but never witnessed my ability.

I then started having a hand to hand combat classes in the rear of the station for any deputy who wanted to receive the training. During one session of training, a Sergeant came back to watch and told me if he got a hold of me, I wouldn't be able to get out. I accepted his challenge, and he got me in a hold from behind. I easily escaped the hold and threw him to the mat and from that day forward he hated me and let me know it. I adopted the saying of, "Never show your sergeant up for any reason."

During a briefing, before going out on patrol, we received notification that the Los Angeles Police expected big trouble from the Watts area of Los Angeles, and if we heard a call for a "999", we should not roll on it until we got further confirmation. The meaning of a "999" is an officer (police) is in dire need of help, and normally, every car in Los Angeles County would respond to the location. However, we were told to remain on patrol until we got official word to proceed to the location because it was about twenty miles to the south along surface roads and it would take quite a while to get there even with red lights and siren. Also, going any distance with red lights and siren is dangerous to others on the road.

I can remember I was riding with a black deputy when the call came in, and I kiddingly asked my partner, "OK Johnson, it's time to choose sides." He didn't think that was at all funny.

The next day we were assigned to proceed to the Watts area and would receive our orders when we got there. We worked twelve hours on and twelve hours off and rode four deputies to a car. It was awful, and I decided right then, I needed to change my career from law enforcement to flying.

Conspiracy

When the Watts Riots were over, I went on the three pm to eleven pm shift. These hours allowed me to travel to the Santa Monica Airport and take flying lessons from a flight instructor named Joe Graham at Santa Monica Flyers. As a kid, I used to fly model airplanes, but I didn't know if I had the skill to fly a real one so, I signed up for a free lesson to see if I could handle it. I flew a Piper Cherokee, which is a single engine, low wing airplane painted yellow and blue. Those colors were the signature of the Santa Monica Flyers flight school. After becoming airborne, the instructor allowed me to take the controls of the airplane, asking if I had flown an airplane before. I told him only model airplanes when I was a kid.

I started flying and took my first lesson on September 9th, 1968. Within two weeks I soloed after only nine hours total flying time. With my evening shift on the Sheriff's Department, I was able to practice two hours every morning and every other day. I got my Private Pilot Certificate in October and started training for my Commercial Pilot Certificate. I had told my Flight Instructor, Joe Graham of my plans to become a flight instructor to build up hours as quickly as possible. He switched me to the right seat so that I could train for my flight instructor rating at the same time I trained for my commercial. It didn't take long until I was able to quit the Sheriff's Department and start teaching students full time at Santa Monica Flyers in February of 1969.

As a beginning flight instructor, we would give thirty-minute demonstration rides free to potential students. I would fly over the Los Angeles freeway system and point out the number of cars jamming the road, especially during rush hour. I would bring their attention to the speed of the car versus the airplane. I sold almost every demonstration I gave enticing them to take flying lessons. If I signed them up, I would get the student. My schedule quickly filled up with students.

I was giving a flight lesson to Vic Petric, who was flying the airplane and we had just departed west off Santa Monica Airport's runway 21. After reaching approximately four hundred feet altitude, the engine quit. I immediately contacted the control tower, took control of the airplane and rolled it into a steep left turn (wingover maneuver) and landed in the opposite direction on runway 3.

My heart was in my throat, and after landing the owner of Santa Monica Flyers, Ms. Betty Miller chewed me out for not slowing the airplane more before touching down. I couldn't believe her for I had gotten the airplane back unscathed, but she made a big thing out of the landing. My first emergency, however not my last.

The mechanic discovered one of the baffles inside the muffler broke loose and covered the exhaust port choking the engine causing it to fail. This same mechanical malfunction occurred to another instructor the previous year when he and his student departed San Fernando Airport and encountered the same engine problem. The student killed and the instructor severely injured.

The government's G.I. bill gave me the money to get my commercial and flight instructor ratings in the helicopter. My instructor, Art Jones who was also an airplane flight instructor, did the same thing Joe Graham did by allowing me to practice everything from the right (instructor's) seat. Thus I was able to take the commercial flight check just as I was ready for my flight instructor check with the Federal Aviation Administration's Aviation Safety Inspector, Mr. Bill Culliton. Mr. Culliton was very thorough and strict but I enjoyed flying with him, he was good. Tests for any Flight Instructor rating had to be with FAA's Aviation Safety Inspector, where the other ratings, private, commercial, etc. went with an FAA-appointed pilot examiner who was not employed by the FAA.

The owner of Santa Monica Flyers, Chuck Miller leased a Bell 47D Model, a helicopter which was used for flight training and was very expensive to rent at $65.00 per hour. Chuck would allow the helicopter instructors, Art Jones and me to fly the helicopter for one half hour free anytime we cleaned and washed it. I took full advantage of this and flew the aircraft at night to build up flight hours for obtaining my Airline Transport Pilot rating in the helicopter.

I washed the helicopter on a regular basis, and one of the free flights took my wife Jeanne for a ride and landed in the backyard of my parents' house and had lunch. We waited with bated breath for any complaints from the neighbors, but none came.

The Los Angeles Times was a regular customer renting the helicopter to take aerial photographs. During the East LA riots, I flew the helicopter with George Fry, an LA Times photographer. He took photos of the Los Angeles Sheriff's Department personnel using scrimmage line tactics in an attempt to control the riotous crowds. With my experience as a Sheriff's Deputy during the Watts Riots which I considered total mayhem, the police displayed excellent control and soon had the crowds subdued. I was thoroughly impressed.

The helicopter proved to be an excellent tool for the Los Angeles Times and police work. Both the Los Angeles Police Department and the Sheriff's Department used Art Jones and me to train their police officers in helicopters. We established a reputation among helicopter pilots as being two of the best in the Los Angeles area. It was a great honor, and we talked about it often.

Chuck and Betty Miller sold their flight school to a company from Chicago called Lease-A-Plane. When the new company took over, they brought in various makes and models of new airplanes giving us a variety of aircraft to fly. Chuck had nine single-engine Cherokees and one twin-engine Piper Apache which was very underpowered and didn't have a very good cruise speed. Kiddingly, we would refer to this aircraft that if one engine failed you would use the good engine to guide the airplane to its accident site.

Lease-A-Plane kept the same FAA Approved flight courses and promoted me to Chief Pilot and Chief Instructor for all flight training. Art Jones made Chief Pilot of the Air Carrier operation in both the airplane and helicopter. I continued as chief pilot for about six months until the Federal Aviation Administration hired me to become an Air Traffic Controller at the Santa Monica Tower. I moved from one side of the airport to the other in the control tower.

I received my training at the FAA's Aeronautical Center in Oklahoma City, Oklahoma. Training lasted five weeks, and I graduated second in my class. When I returned to Santa Monica Tower, I received on the job training which lasted six months. The Tower Chief, Larry Morton asked what I wanted to do with the FAA. I told him I wanted to become an Aviation Safety Inspector with FAA Flight Standards which is in a different field than Air Traffic. Larry got upset with me, and he told me he would never recommend me for that job because the FAA has spent a lot of money training me and he was not about to let me go. When he finished his promotional speech, I handed him my two weeks' notice and left because I was offered the Chief Pilots Position at Lease-A-Plane again on the other side of the airport.

Chapter 5

JACQUE COUSTEAU

In June of 1971, right after reporting to Lease-A-Plane as Chief Pilot, I was asked if I wanted to accept a contract to provide aerial photography in a helicopter to Jacque Cousteau on his Calypso. His ship, anchored at the Blue Hole in Belize, British Honduras. The manager of Lease-a-Plane, Ken Krueger and I reviewed every option available of getting the helicopter from Santa Monica to Belize. We knew we had to install a float system (pontoons) on the helicopter because the Blue Hole was located approximately seventy miles offshore into the Gulf of Mexico. Transporting a helicopter on a trailer with pontoons was impractical because it would be very unstable. For stability, the helicopter needed to be on the skids while on the trailer.

Another problem facing us was the limited distance the helicopter could fly with its limited fuel system. The normal fuel system carried enough fuel for approximately two and a half hours. We needed about four hours with floats creating drag. The floats created a lot of drag limiting our airspeed to approximately 70 miles per hour. The added flying time would give us the range we needed to go from one city with approved fueling facilities to the next. In between these fueling facilities, we couldn't count on any fuel being available other than maybe finding a location that had it in fifty-gallon drums. Fuel in drums would probably be contaminated so it would need to be strained through a chamois to remove any water or foreign matter.

We reached a feasible solution. We would leave the skids in place while we trailered the helicopter behind a rented pickup truck from Santa Monica Airport to Brownsville, Texas. There we would install the pontoons and fly the helicopter to British Honduras. To satisfy the fuel problem, we installed small brackets on the outside of the helicopter's frame that supported four, six-gallon fuel cans tied to the structure. There was also two litter baskets place and secured on each landing strut giving us additional storage space. Before our departure, we received a telegram from Mexico City to take with us for clearing Mexican Customs. The last time I flew a Helicopter into Mexico the customs people wanted bribery money to not ask questions concerning our purpose for being in Mexico.

Walter Wilson volunteered to be a passenger aboard the helicopter I was flying and support me while I flew the helicopter to British Honduras which was perfectly legal. Walter was a rated airplane pilot but not in helicopters.

We left for Brownsville, Texas early in the morning and things went rather smoothly until we crossed the Texas state line.

The Local Police stopped us and said we needed a Bond to protect the Texas Highway in case our trailer crashed onto a Texas road. He showed me the state ordinance which required a bond for any unusual vehicle traveling in Texas. So, we went to a bond office in the local town and purchased this bond. The bondsman was the brother-in-law of the local Chief of Police.

Traveling in Texas from El Paso to Brownsville took us two full days. We arrived late in the afternoon and decided to park the truck and helicopter at the airport and get a good night sleep before installing the pontoons. The skids of the helicopter had removable wheels which allowed us to roll the helicopter into a hanger for the night. In the morning we positioned the helicopter under a large hoist which lifted it clear of the ground enabling us to install the pontoons. On top of the rotor mast which is the balancing point for a helicopter, is a round metal loop about three inches in diameter. We connected the hook from the winch through the loop and lifted the helicopter off the ground.

After we installed the pontoons, another problem appeared. We had no way of rolling the helicopter out the hangar door. Our only alternative was to fly the helicopter out of the hanger. We talked to the manager of the hangar and explained our situation. He permitted us, and we advised him to cover everything that could get damaged by dust and dirt stirred up by the rotor downwash. Clearing the top of the hangar's door opening appeared to be the biggest obstacle. I started the engine and increased lift with the collective pitch[1] until about six inches existed between the bottom of the pontoons and the hangar floor. Dust was going everywhere, and the hanger manager started to have second thoughts. I had to hurry, so I started moving toward the open door with about two feet separating the top of the door and the rotor system. I increase my forward motion and flew the helicopter through the door and set it down outside. We spent the next two hours loading all our gear, fueling the gas tanks and cans, and checking all the systems. We lifted off and flew to Matamoros, Mexico.

FIGURE 11 - MEXICAN CUSTOMS AT THE HELICOPTER

When we landed the first thing we did was present that telegram to the customs people. Their reaction was one of excitement and wanting to help. Fortunately, we had no problem with the language because most of the customs people spoke English. I don't know what was in that telegram, but whatever it was it removed any apprehension I had about entering Mexico.

---

[1] The Collective Pitch is a lever located alongside the pilot and has a twist throttle like a motorcycle on the end. When you raise or lower it the rotor system's lift is increased or decreased making the helicopter go up or down.

Conspiracy

The customs people wanted their pictures taken alongside the helicopter, and the Commandant of the Airport came out to send us off. I've been to Mexico many times, and this is the first time I ever experienced anything like this.

The helicopter had no navigational instruments on board. The only thing we had to navigate with was aviation charts. Fortunately, until we get to Villahermosa, we wouldn't be far from the Gulf Coastline. From Villahermosa to Chetumal it was steering with a magnetic compass that bounced around and shifted forty-five degrees due to the vibration of the helicopter. This part of the trip had to be done by dead reckoning meaning by reference to ground obstacles only because the compass was unreliable.

We flew south along the coast planning on spending the night in Tampico which was two hundred and eighty-seven miles from Matamoros. Traveling at about sixty-five to seventy miles an hour, it took us about four hours and twenty minutes to get there. This Bell 47G only flew about eighty miles an hour, and with the drag of the pontoons it slowed us down about ten miles per hour. After about an hour and a half from Tampico, we needed to stop and refuel the main tanks. We circled a spot that looked as though there wasn't anyone within a hundred miles. However, when we landed and before the rotor blades stopped, the helicopter was surrounded by the local populace. Fortunately, they were very friendly and excited to see and be near a helicopter.

That part of Mexico didn't have very good medical facilities available to the people. Some of them had deformities that were obviously unattended by doctors. We saw a few of them with cleft palates that went unattended. Their faces were so distorted that it was sad to see.

Walter and I poured the fuel into the tanks as fast as we could, started the helicopter and got the heck out of there. Tampico was a fairly large city with a very modern airport. The airport officials again treated us like royalty. We refueled and were airborne in about one and a half hours.

We arrived in Vera Cruz two hours later, and again we were treated well after the officials read that telegram. The airport officials had the same manner about them as in Matamoros. Excited to see us and they couldn't do enough for us. We were very grateful the airport personnel put the helicopter inside a hanger, and we went to the hotel for a well-deserved rest.

The city had a square right in the middle of town where people gathered. It was very pleasant, and the food was nothing less than outstanding.

The sun had barely risen when we were on our way to Villahermosa, two hundred seventy-one miles to the south. After flying for about two hours, we landed in the mountains and refueled in the wilderness. This time no one came to watch, and we felt pretty secure.

FIGURE 12 - LAGUNA DE TERMINOS

After two more hours of flying, we landed at Villahermosa. We ate and refueled all the fuel tanks without crowds of people watching. The airport officials were also very friendly after reading that telegram. I hate to imagine what we would have gone through without that piece of paper.

Leaving Villahermosa, we traveled northeast eighty-four miles to Ciudad Del Carmen right on the northern shore of the Yucatan Penninsula. This city was right on the water of a huge inlet called Laguna de Terminos. The local beach had Coconut Trees all along the shore with a lighthouse at the end. The strip of land was only about three miles wide from the Gulf of Mexico to the inlet. It was obviously a tourist town because it had many hotels and everything was well manicured, and the surroundings were gorgeous.

Refueling, we continued across the inlet using our only means of navigation the compass and dead reckoning. We couldn't see the other side of the inlet. So to navigate, I picked a reference point in the distance and tried to fly straight at it. We headed toward a road we would follow across the

Yucatan Peninsula and on into Ciudad Del Chetumal, some two hundred and thirty-two miles of nothing but jungles.

Walter and I thought we could make Chetumal before it got too dark to fly but we were wrong. It was beginning to get dark, so I decided to land in a clearing in the jungle before it got too dark to be able to see. We landed near the Campeche Inca Ruins. I would have loved to explore the temples, but there was no way I would walk through that jungle at night. As it got progressively darker, the noises coming from the jungle grew louder.

We had a litter basket strapped to the skids in which Walter slept while I opened my sleeping bag and pulled it around me in the cockpit. I wanted no part in sleeping on the ground. The noise coming from the jungle was scary. We heard cats screaming and growling, birds making all kinds of noises along with sounds from other animals or whatever. That night I confessed all my sins and promised God I would do better. It was a long night, and with the sleeping bag over me trying to keep the bugs from attacking, I perspired like crazy. When I removed the sleeping bag to get air, I was attacked by various flying insects. It was not fun.

Morning finally came, and we were up at the first crack of sunlight headed for Chetumal. We were following a stream or river that had something very large in it that surfaced and then sank back down. We didn't know what they were, but my guess was Manatees.

FIGURE 13 - LEAVING MEXICO AT CHETUMAL

The airport at Chetumal was on the edge of the jungle and had some various tropical trees all around the runway and terminal. It was beautiful. I'm sure the thought of coming back into civilization after spending the night in the Yucatan Jungles helped us to see the beauty better.

Alongside the runway was a World War II B-25 Mitchell Bomber just sitting and rotting away. The airport commandant told us it landed one day and the flight crew caught a cab, left and never returned, and that was about two years ago. It appeared it was from the filming of "Catch 22". I'm sure the airplane collectors have since salvaged that airplane by now. We climbed into it and looked around, and it appeared to be in pretty good shape.

We left on our last leg of the journey to Belize, British Honduras. The east part of the country was thick with jungles, so we flew along the coast for one hundred and sixteen miles and landed at the Belize Airport. Waiting there to greet us was Jacque Cousteau, Ken Krueger, my boss who had flown there in a Cessna 172. I was jealous. He could fly while relaxing in an airplane while I had to fly most of the time tensed up in the helicopter. I was still tense thinking about sleeping in the Yucatan Jungles the night before.

THE BLUE HOLE

Cousteau told me we were a day late and they were considering sending a search party out to try and locate us. I assured him we rushed as much as we could but got hung up in Brownsville installing the pontoons and then flying it out of the hanger. He laughed. This lens used for this picture is called a "Fisheye."

Conspiracy

FIGURE 14 - THE CALYPSO INSIDE THE BLUE HOLE

After lunch, Ken climbed into the helicopter with Walt and me and headed away from Belize toward Turneffe Islands approximately forty miles offshore into the Caribbean Sea. Turneffe Islands is a reef twenty-two miles long and eight miles wide. In the center is a circular reef where the ocean floor had fallen some four hundred feet. Falco, Captain of the Mini Sub, determined the depth on one of his dives. The color of the water throughout these islands is a turquoise blue and the color of the water inside the reef surrounding the Blue Hole is a deep purple. It is really beautiful. There is a small island just south of the Blue Hole with a lighthouse. On this island is a disserted house with no windows. Cousteau had it set up for us to use as our sleeping quarters and piled four cases of beer in case we got thirsty. The first night we stayed in this house but didn't even try to drink any beer. After that night in the Yucatan, we just wanted to sleep.

FIGURE 15 - THE CALYPSO ANCHORED IN THE BLUE HOLE

The following morning we met the Light House Keeper. He was an ex-Belize Police Officer who left his job to care for the lighthouse. He said he was much happier and made more money fishing than he did as a Police Officer. He warned us to not go to the far end of the island because scorpions and tarantula spiders overran the area.

The next day I flew the helicopter with Walt and Ken to where the Calypso is setting inside the Blue Hole. We landed on the water near the reef and tied the helicopter to a rock where the crew had a rope attached. We climbed into a rubber launch with an outboard motor that took us to the Calypso. We met Bob Dill, from the San Diego Marine Laboratories, who at the time had the deepest record in a diving bell at 43,000 feet. Andrea Labone, Production Manager, Pierre Captain of the Calypso, and Jacque Renoir, photographer and great-grandson of the famous painter Renoir. We all sat at a table and talked about what they expected of us.

The captain was going to take the Calypso out of the Blue Hole to a point where they first entered. I was to fly over them with a photographer strapped to the outside of the helicopter on one of the pontoons and film the Calypso coming down the channel and into the Blue Hole. The photographer gave me instructions on how he wanted the helicopter positioned as the Calypso went under us. I agreed, and we all left to accomplish our assignments.

Ken sat on his butt and watched comfortably from the Calypso.

FIGURE 16 - KEN AND SHARK JAW & LOCALS REFUELING THE HELICOPTER ON THE ISLAND

I decided to return to the island and fill the helicopter with fuel allowing us to complete the filming of the Calypso's return to the Blue Hole without having to stop for fuel and delay the filming. I had never flown a helicopter with floats before and consequently was unprepared when it uncontrollably turned one full turn to the left because of the engine's torque. Fortunately, the tail boom was unobstructed and didn't hit anything when it turned.

With the photographer strapped to the outside of the helicopter, we lifted off and flew back to the island and refueled. When we were leaving to go back to the Calypso with the photographer strapped to the outside, I attempted to lift off, but the film magazine for the camera had slipped unnoticed into the path of the collective jamming it and limiting its travel. Without having full movement of the collective, we descended, and the front of the pontoons struck the water causing the nose to go down. We were on the verge of crashing with the rotor blades hitting the water when I pulled the control stick all the way to the rear while jerking the collective upward with all my might. This abrupt action dislodged the film magazine, and the nose popped out of the water. We returned to the island and cleaned the water off everything especially the camera and its lens. We also had to calm ourselves after that mishap.

When we returned to the Calypso, the captain had repositioned it just outside the reef. We radioed what happened and that we were ready. They pulled anchor and began their return to the Blue Hole. They proceeded down the channel, and I flew and positioned us directly the front of the ship's path.

As they came close, I raised the helicopter to above the Calypso, and they sailed under us with the photographer filming their progress. We repeated this maneuver several times until the Calypso was back in the Blue Hole.

I landed the helicopter on the water near the side of the ship, and a launch came and picked us up. During lunch, the wind picked up causing small waves that rocked the helicopter fore and aft making the rotor head strike the mast. We all became worried something was going to get damaged which would ruin the rest of the shooting and put us out of business.

The machine shop cut two pieces of metal approximately two feet wide and four feet long. They bent them in the middle to form a "V." Two divers tied these two pieces of metal at both ends of the helicopter fuselage and the resistance of these metal slabs almost totally reduced the movement of the helicopter to zero. It was amazing, and the problem no longer existed. One crew member called these plates "Flopper Stoppers."

FIGURE 17 - THE HELICOPTER AFTER INSTALLING FLOPPER-STOPPERS

Walt tied the rotor system down with a pair of two wooden blocks designed specifically for a helicopter rotor blade. Each block contained a curved side matching the curvature of the rotor blade. What we didn't notice was that there was a deformity on one block that when tightened around the rotor blade, it put a small one-inch dent breaking the seal on the trailing edge of one blade. The blades are wood, and if any water got into this damaged area, the wood would swell and ruin the entire rotor blade.

Conspiracy

The machine shop mixed epoxy and smoothly spread it into the damaged section of the blade making it waterproof and smoothing it to match the curvature of the rotor blade. It worked perfectly.

Someone had radioed this information to Belize, and the airport officials wanted us to return to Belize to file an accident report but Cousteau convince them it was nothing, and it was forgotten. We let the epoxy cure before flying again, so we spent the night on the Calypso.

Andrea Labone, with his thick French accent, told me "Your launch is waiting." I thought he said your lunch is waiting. So I replied, "No thank you, I've already eaten." He was referring to the small boat alongside the Calypso. I felt like a fool, but everyone who heard us broke out laughing.

That night, one of the crew had a pet Marsupial. It was very friendly and enjoyed the attention. That night we were all sleeping on deck when the Marsupial came up to Ken while he was asleep and stuck its tongue in his ear. Its tongue was long, and Ken came up out of his sleeping bag like a bolt and scared the poor animal to death. In a defensive move, it started attacking Ken with a vengeance. It started scratching and biting at Ken who had to crawl inside the sleeping bag for protection. It woke all of us and when we found out what happened, we started laughing and couldn't stop. The owner of the animal came and took it to bed with him, and we were eventually able to fall asleep again. At the bow (front) of the Calypso was a hatch covering a shaft that went down to an observation deck that was below the water level. You could observe divers in the area working while under water. I spent some time watching the divers work underwater and the various fish swimming around them.

FIGURE 18 - UNDERWATER SHOT FROM OBSERVATION DECK AND INSIDE THE OBSERVATION DECK

The Calypso crew worked from sun up to sun down with very few breaks. I kept pace but flying the helicopter under the stress of maneuvering it to capture aerial photographs caused me to tire early. But I held on and kept up with everyone. It was no easy chore.

This grueling schedule went on for four days. Each night we tied the helicopter down with the "Flopper Stoppers" which made us feel secure in the hope it wouldn't get damaged should the wind start again.

The next to last day on the Calypso we were all asked if we wanted to go snorkeling to get lobster for bouillabaisse soup. A diving platform installed on the stern (back) of the ship is where we would enter the water. I was first to climb down and stand on it's the platform. Before entering the water, I looked down, and under the platform, a Barracuda that looked like it was around five feet in length was just floating. I refused to enter the water and Falco (Mini Sub Captain) laughed at me and chased the fish away. By this time everyone was there and ready to jump into the water. At the time none of us were certified divers, so we were only allowed to snorkel. We all agreed to get our diver certification when we got back to Santa Monica.

FIGURE 19 - FALCO AND SUPPORT TEAM

We entered the water and swam to the reef on the edge of the Blue Hole and started looking for lobster located in holes among the rocks. I spent my time watching barracudas swim up to me looking me over. I could see them smacking their lips, but nothing happened. I finally got courage and swam over to look into some holes around the coral. Many of the holes had sea

Conspiracy

urchins blocking them so; we couldn't reach in to grab a lobster. A sea urchin has a round black body with numerous spines of various lengths sticking out all over and if you got stuck by one you had to be concerned the wound would get infected. If the end of the spine broke off in your skin, it would be very painful. We stayed away from those creatures.

We gathered a few lobsters and swam back to the ship and gave them to the cook. That evening we had bouillabaisse soup. It was out of this world.

Walt, Ken and I flew back to Belize in the helicopter and spent the night in town. We had dinner and drinks at a local bar. The temperature in Belize was such that most establishments didn't have windows, only large overhangs that prevented rain from coming in. The humidity was extremely high and very uncomfortable.

At the back of this restaurant was an open window overlooking the canal. In the canal were hundreds of catfish. If you threw anything into the canal, the fish would go after it. It looked like a school of Paraña attacking a bloody animal. I hate to think what it would look like if a human were to fall in. Fortunately, catfish don't have the teeth of a Paraña.

The next morning we all left for home. Ken flew ahead of us on the same route as we flew coming down. We stopped at Ciudad Del Carmen and spent the night at a resort. It was unbelievably beautiful.

FIGURE 20 - KEN GETTING READY TO LEAVE BELIZE

Early the next morning we left for Brownsville, Texas. I flew that helicopter as fast as it would allow and we reached Brownsville that evening even though we had to refuel every couple of hours. We called Matamoras Air Traffic Control and asked if we could continue to Brownsville without stopping for customs because we hadn't bought anything and had just enough fuel to make it to Brownsville if we didn't stop. Matamoras approach control permitted us, and we received authorization from the US Air Traffic Control, and we landed uneventfully at Brownsville.

Two women came up to us as we were unloading and started talking about where we had been. We had enough fuel so I asked Walt to stay on the ground and I took both of them up for a joy ride that lasted about ten minutes.

After landing, we went to the same hanger as before, and the owner provided us with a trailer to roll the helicopter through the door and into the hangar. The only drawback was, there were no means of getting the helicopter onto the trailer other than flying it on. I flew the helicopter and put it into a hover as smoothly as possible and landed it on the trailer. Then I gently set it down on the two metal slides built for metal skids, not pontoons. I reduced the throttle to an idle and let it cool before shutting the engine off. As the engine RPM reduced to zero, the rotor system started slowing down also and what concerned me was that the slower the blades turned, the more unstable it would become. Fortunately, it remained steady while the engine cooled and I was able to shut it down after the one-minute cooling requirement.

Once the helicopter's rotor blades stopped turning, we began removing the pontoons. The hanger manager said he has never seen anyone hover a helicopter as steady as I did. I felt honored.

Walt and I deflated and removed the pontoons, reinstalled the skids, and removed the rotor blades from the hub, and laid them alongside the skids on the trailer. We departed Brownsville that evening and started our trip back to Santa Monica taking the two women with us. They lived in Los Angeles and had flown to Brownsville to visit family.

We drove day and night until we were back in Santa Monica. The next two days we relaxed and caught up on our sleep. The girls didn't live far from Santa Monica, and after unhooking the trailer, we took them home.

Two weeks later, my wife and I received an engraved invitation from Jacque Cousteau. It invited us as his guests to attend Andrea Lebone's art exhibition at a bank in San Diego. The exhibition was to view Andrea's paintings that he created while underwater. I called Kenny, and he had received an invitation also. That Saturday night Ken, his girlfriend along with me and my wife Jeanne flew a single-engine airplane, a Cessna 182 to San Diego.

When we arrived at the bank, the guard took our invitations and escorted us to the area reserved for Andrea and his paintings. As we entered, Cousteau saw us and excused himself from whomever he was talking to, grabbed two glasses of Champagne, walked to where we were standing and gave the glasses to each of the girls. We introduced him to the women and at his request, began telling him about our trip home. He listened intently. Jacque Cousteau was a remarkable man and gentlemen and who were we? A pilot, and a manager of an aviation flight school, yet he respected us, and we got along superbly.

After the Art Exhibition was over, Cousteau invited us to dinner at a restaurant in the San Diego Airport. During dinner and drinking champagne Ken received a telephone call and said he had to get back to Santa Monica immediately. Excusing ourselves, we promptly left for the airport and flew back to Santa Monica. The weather had turned overcast, with light drizzle and fog. Ken said he was too drunk to fly and so I got elected. I didn't feel drunk, but that didn't matter because I was the only one available to fly back to Santa Monica.

When airborne, Ken and the two girls fell asleep leaving me along with flying the airplane in bad weather back to Santa Monica. Nice friends.

Afterward, we were invited to various parties with members of Cousteau's Society and started socializing with them on a regular basis. Ken and I went to numerous parties at their homes or apartments. One such occasion, we were at one of Cousteau's crew's apartment, and three of them had gone diving picking lobsters from poacher's traps. They didn't feel bad about what they were doing because the traps were illegal anyway.

The next week I was contacted by the manager of the local FAA, Flight Standards District Office and asked if I would be interested in becoming an FAA Aviation Safety Inspector. I told him "Yes," and because I was a past FAA Air Traffic Controller, I did not have to go through the tedious process of applying through the normal channels. I could just come aboard directly and in a relatively short time frame. Two weeks to the day, I reported to the FAA's, General Aviation District Office in Fresno, California, as an Aviation Safety Inspector.

Chapter 6

MY CAREER WITH THE FAA

I reported to the FAA's, General Aviation District Office at the Fresno Air Terminal Fresno, California. The office was located at the base of the Air Traffic Control Tower and had a typical government décor; gray desks walls were a light tan and floors pale white.

I started receiving job training with Principal Operations Inspector Jack Patrick. My pay grade was a GS-9 level 7 because I lacked experience as a pilot and the level 7 was because I had Air Traffic Control experience. To my understanding, I Fresno needed a helicopter specialist, but when I arrived, I found that there was a helicopter specialist already there by the name of Joe Pyper. He had about seventeen thousand hours in helicopters; I had around eight hundred hours. He was so much more qualified than me; I was not disappointed. It didn't matter to me though because I had more hours in airplanes than helicopters.

My first week on the job, I was thoroughly tested both in airplanes and helicopters. Jack rented a twin-engine airplane called a Beechcraft Baron. We flew all over the San Joaquin Valley to Bakersfield, Porterville, Visalia, etc. We practiced instrument approaches where ever there was an instrument approach we could use.

Then Joe Pyper took me out to the Porterville airport where we rented a Bell 47 D helicopter. The helicopter lacked hydraulic boosted controls which is what I was used to, and it felt like I was flying a Mac Truck. Hydraulic controls are like power steering. I did OK I guess because I continued working.

Conspiracy

While in the office, Jack had me reading all the FAA Orders, Manuals for Air Carrier, Agricultural, including accident and incident Handbooks. I was scheduled to attend the FAA Training Academy as soon as possible, but due to FAA's budget constraints, I wouldn't be attending the academy for some time.

I hate reading, and it was total boredom for me, but I got through them all, and then Jack had me reading them again. Additionally, Jack had me read regulations, and I thought there were only two or three, but I found there were many more. If you can't sleep at night, read regulations and sleep came naturally and very quickly.

Within a month, my wife Jeanne got a transfer with the Pacific Bell (Telephone Co) and moved up to Fresno with me. We rented a house for the first year to learn the area before buying our own house.

Bakersfield, California was our busiest airport, and we visited there about once or twice a month. We stayed in the hotel at Meadows Field and visited companies that made their living operating airplanes or helicopters. Jack introduced me to most of them, but we were there for only two days, so we had to start with only a few at a time.

The FAA offices in Fresno were small, and there were only four operations (pilot type) inspectors and three airworthiness (Mechanics) inspectors. I shared an office with Joe Pyper. Joe's hobby was Square Dancing, and when he traveled up north, he would visit various dance clubs.

The northern area of our territory was on both sides of the San Joaquin Valley as far north as Stockton, California. Merced and Modesto were their busiest airports, and the Principal Inspector in charge was Burt Rhodes.

Eventually, Joe told me he was in World War II as a Sea Going Marine aboard the Battle Ship, USS Arizona on December 7th, 1941, when the Japanese attacked Pearl Harbor. He was wounded, blown overboard and watched the entire attack from the middle of Pearl Harbor. He was subsequently brought back to the US as a hero and asked what he wanted to do. He told his superiors he wanted to fly. So they sent him to Pensacola, Florida where he entered naval flight training.

Joe didn't do so well at first, but they weren't about to let him fail. When he ground looped and damaged a Boeing Stearman, A75N, which was a WWII double wing, open cockpit primary pilot trainer, they told Joe not to worry they would get him another airplane. He graduated from flight school and flew a Corsair F4U off an aircraft carrier in the Pacific. That was no easy task. When the US first tried to use the Corsair F4U on Aircraft Carrier operations, they had numerous accidents because the long nose on the aircraft made it difficult to see well enough to land. These aircraft went to the British, and they mastered a landing technique.

After the war was over and the Korean War started, the Marine Corps asked Joe if he wanted to fly helicopters. Joe said yes, and met his niche. He turned into one of the best helicopter pilots in the Marine Corps and eventually the FAA.

Two years after I arrived in Fresno, Joe Pyper transferred to Anchorage, Alaska. He died when the airplane he was in as a passenger hit a mountain while attempting an instrument approach into a remote area during bad weather. His family had him cremated, and his ashes sent to Honolulu, Hawaii. A Military Diver took his ashes and buried them inside the hull of the USS Arizona. All survivors of the USS Arizona on that fateful day of December 7th, 1941 are allowed this honor. The deceased returns home to rest with their shipmates.

Enforcement of the aviation regulations is a primary function of an Aviation Safety Inspector along with accident investigation. These two primary functions were hand and hand because many times one will lead to the other. My enforcement cases were numerous, but I always tried to get compliance from the operators and pilots through education, and understanding and not enforcement.

We conducted seminars, pilot education programs and re-examination of pilot skills. Anytime a pilot wanted to have an FAA inspector evaluate his/her pilot skills they could schedule an evaluation with one of the inspectors. Normally, the Accident Prevention Inspector (APS) does all the evaluations. He/she would do all the pilot evaluations. However, if during one of these evaluations should the APS detect a deficiency that individual had immunity from losing his pilot certificate and the evaluation would continue until the deficiency was corrected.

On two occasions I received letters of commendation from the Regional Headquarters in Los Angeles for outstanding enforcement reports.

Accident Investigation is also one of the primary responsibilities in the FAA. However, the National Transportation Safety Board has primary responsibility, and they delegate certain categories of accidents to the FAA. As an example, an aircraft sustains minor damage with no injuries the investigation would be conducted solely by the FAA.

The FAA investigates all experimental (home built) aircraft accidents. An accident that is not an experimental aircraft and where someone receives fatal injuries the NTSB is the primary investigating authority.

There are two NTSB offices in the Western Region of the FAA. One located in San Francisco, and the other in Los Angeles. The NTSB investigator had to travel approximately four hours to our district when there was a fatal accident. The same driving time existed whether they traveled from San Francisco or Los Angeles.

I got so proficient in investigating accidents, the NTSB would travel to our district, and I would hand them a completed accident report, and a roll of film and they would return home. I was asked numerous times to come to work for the NTSB as an investigator, but I decline. I enjoyed the diversity of assignments working with the FAA, not just accident investigation.

In September of 1974, I received a telephone call from Cape Cod, Massachusetts informing me that my father, while on vacation, had suffered a heart attack and had walked into the hospital. They had been visiting his sister Audrey when his attack had occurred.

According to my Aunt Audrey, my family has been in the Massachusetts and New York area since they arrived in this country on the Mayflower.

My brother and I boarded a United Flight to New York City. From there we bought tickets on a DC-3 to the Hyannis Cape Cod Airport. It was at night, and one of the pilots would shine a flashlight on one of the engine cowls checking it all over. My brother seeing this made a loud statement, "No oil is good oil." People who heard him started laughing. All of a sudden the cockpit door came open, and it was obvious the door lock didn't work. Federal Aviation Regulations require all cockpit doors must be locked to keep anyone from entering the cockpit. So to keep with the regulations and because my seat was in front of the cockpit door, I put my foot on the door and kept it shut. An hour after leaving New York's Kennedy Airport, we landed at Hyannis.

My father was in the local hospital but his younger sister, Audrey was hesitant to enter dad's room because it was same room her husband Clifford, had died in three years earlier.

My brother and I spent a few days at Cape Cod visiting my dad but decided to return to California because we didn't want him to think we were there waiting for him to die. My mom could call us if we were needed.

After being home for a couple of days, I received a call from the Cape informing me my father had passed away. I was devastated and called my brother. He wanted to split the expense to rent an airplane and fly to Cape Code. We rented a single-engine airplane called a Bellanca from an operator in Van Nuys, California and flew to the Cape.

When the weather was bad, I would fly because at the time my brother didn't have an instrument rating. Flying to Albuquerque, New Mexico my brother Buddy was at the controls and because of incoming weather; he kept getting lower and lower trying to stay clear of the clouds. We were on a flight plan that didn't allow us to fly in the clouds and radio contact was restricted because of the mountains. If we inadvertently entered the clouds, I was prepared to take over and climb until clear of all weather. To do this, however, I would have to call Air Traffic Control and get a new clearance allowing us to fly into clouds. However, we were able to stay clear of all clouds and landed at Albuquerque for fuel and food.

From Albuquerque east, the weather was clear and beautiful. The airplane we were flying was a Bellanca Viking which had a cruising speed of around one hundred and seventy-five miles per hour and had excellent radio and navigational equipment. It was autumn, and the leaves and bushes were changing color. The various colors of the trees and bushes were spectacular.

My brother and I made all the arrangements and held the funeral at the local funeral parlor. I demanded a closed casket and received no argument. I reasoned that I wanted to remember seeing my father alive not in a casket and to this day I am thankful I made this request.

My mother and father lived in a mobile home in Bodfish, California located thirty-two miles east of Bakersfield in the southern Sierra Nevada Mountains. They took this vacation to visit his little sister in West Barnstable, Cape Cod, Massachusetts. They traveled in their International Travel All towing a twenty-five travel trailer. During the time my father was in the hospital, my mother rented a campsite at a local trailer park. After the funeral, my brother and I flew the airplane back to California while my mother flew back to Los Angeles to be with my brother's wife, Kitty.

After we delivered the rental airplane to Van Nuys, California, I flew up to Fresno met Jeanne, and we both flew back to Cape Code to bring the Travel All and trailer back to California.

We had my father's remains cremated, and my aunt took his ashes and buried them next to her husband.

TRANSFER TO OKLAHOMA CITY

I spent four and a half years at the Fresno General Aviation District Office when I received a promotion and transferred to the FAA's Aeronautical Center in Oklahoma City as an instructor teaching helicopters. I also received an Outstanding Achievement Award when our office won National Office of the Year Award. However, I had already left for the FAA Academy when the team from Washington Headquarters arrived for the ceremony.

One of my achievements was taking a picture and having it accepted by the Smithsonian Institute in the Aviation Department. I received a letter of appreciation from them.

Jeanne and I were having difficulties, and it came time for us to go our separate ways. The marriage lasted a little over ten years, and we never argued or had a crossword with one another. Leaving her was one of the biggest mistakes I have ever made in my life. To this day I don't think a day goes by I don't think of her. I have never stopped loving her and I will never forgive myself for leaving.

To become an instructor at the FAA Academy the perquisites were, to complete two courses. Basic Instructor Training Course taught at the FAA Academy, and Management training at Cameron University in Lawton, Oklahoma.

A student in the Basic Instructor Training class by the name of Ray Castro recognized me from Karamursel, Turkey. Ray told me his roommate took judo lessons from me, and after about two months of training, he thought he was the toughest guy around. He got into a fight and lost badly.

When we completed this course, we immediately traveled to Lawton Oklahoma and attended Cameron University for Management training. Some of the people attending this course had personal problems because it was back to school sharing everything, showers, toilet facilities, etc. The only privacy we enjoyed was our private rooms which had only one bed.

After satisfactorily completing both courses, the Academy enrolled me in a helicopter recurrent training course. The primary objective was for me to become familiar with the two helicopters I would be teaching in. One helicopter was the Bell 47G3B1 with a turbocharged gas engine and the other a turbine powered Bell Jet Ranger 206A.

The FAA had their airplane courses at the Will Rogers Airport and the Helicopter Course at the Wiley Post Airport located north Oklahoma

City. The flying portion of the helicopter course we flew from Willey Post Airport to our practice area at the Cimarron Airport now called Clarence E. Page Airport. To simulate landing on a building rooftop was a twenty-foot square concrete platform held up by steel girders twenty feet high located at the southeast corner of the airport.

Just north of the platform, In the middle of the airport was a grass area approximately two hundred foot square with a white painted tire in the center. The tire was a target during autorotation (engine failure) practice. We would fly the helicopter approximately one thousand feet above the ground and bring the engine to an idle. The helicopter would glide, and the rotor blades would aerodynamically keep turning allowing us to glide down to the ground. We would terminate the glide either by landing without the engine or we would terminate the glide in a power recovery with the engine running. A successful autorotation would terminate with the nose of the helicopter directly over the center of the tire either in a hover (power recovery) or on the ground for a touchdown.

Occasionally we would execute an autorotation from ten thousand feet of altitude which was approximately eight thousand feet above the ground and terminate the descent on the ground with the nose of the helicopter directly over the tire. There were other types of autorotation maneuvers taught from off the platform to the ground, and from a hover.

Also taught were External Loads which is where different loads are attached to the belly of the helicopter using a hook. Then we would maneuver the helicopter with various objects under the fuselage allowing the student to experience the various flight characteristics. Examples are: Setting poles, hauling empty barrels in a net, or slinging one barrel half full of water. On more than one occasion, the hook under the helicopter would inadvertently open, and the load descended to the ground and broke into a million, and one piece or the net with the barrels would let loose, and the barrels would no longer hold water.

While teaching an external load operation, my student and I had a telephone pole hanging vertical getting ready to set one end into a hole in the ground. Bob Barton, one of the other instructors, pushed the telephone pole and it started swinging perpendicular to the fuselage. The weight of the pole would make the helicopter rock back and forth and could get out of control if you didn't apply corrective control action promptly. You had to apply control to make the helicopter move in the same direction as the pole causing the pole to stabilize directly under the helicopter.

The FAA inspectors receiving the training at one of our courses often evaluated our instruction as second to none comparing us with other helicopter flight schools. The military couldn't compare to the training received at the FAA Academy. It was rated number one overall. All the instructors were aware of this honor, bestowed upon us, and we all worked hard to keep it.

Two years I taught helicopter training and then I advanced to jet airplanes. The managers in charge of training felt helicopter pilots could only handle an aircraft that traveled one hundred and twenty miles an hour. Jets were around five hundred miles per hour, and they felt helicopter pilots could not cope with this type of speed.

There was a helicopter instructor by the name of John Paulson who transitioned from the helicopter to a twin-engine jet called a Sabreliner. He was a great pilot and did a fantastic job flying. He led the way for other helicopter pilots to transition to jets and I kept alive what he started.

The North American Jet Commander was my first jet airplane. I taught in that for a year and then transitioned to the Sabreliner, then to the Learjet. I encountered a few emergencies in the Sabreliner. On two occasions, I lost hydraulic pressure and had to plan the approach to Oklahoma City Airport using emergency braking. When you utilize emergency brakes, you pull a red handle in the center pedestal which diverts hydraulic fluid from the main source to a separate smaller tank. Then you had to pump the brake pedals to build up brake pressure. It's like a car when you get air in the lines. You pump the brake pedal to raise the pressure in the brake lines, and you have partial brakes. A lot of planning is needed to accomplish this maneuver, so you need a long runway.

The Academy had a program for instructor pilots to maintain proficiency along with currency. The Academy provided seven and a half hours of piloting a jet airplane per quarter. Provided we taught in that airplane and so long as our co-pilot was also qualified.

When things were slow, we would fill the airplane up with other instructors and fly to New Orleans for seafood or Denver, Colorado for steak and so on. We would be back at the Academy every night though.

One day, John Paulson flew the Lear Jet to New Orleans to pick up fresh shrimp for a party we were having at his house. On the way back the box leaked water, and the airplane smelled like a fish when it arrived back in Oklahoma City. We sprayed shaving lotion, and it only made matters worse. We finally washed the carpets which finally dissipated the smell.

Four and a half years after arriving at the academy I received notification that I was selected for a promotion as a Regional Office Specialist at the Western Region Headquarters in Los Angeles. I was finishing my last class in the Learjet and was on short final to Will Rogers World Airport for the last time when I heard the FAA was going to cancel any further Learjet training. My comment was, if they take away my toys, I'm going home.

While at Oklahoma City I met and married Linda Perry. Linda had two girls from a previous marriage close to Travis AFB near Sacramento, California. The girls were Stephanie, seven years old and Melanie, two years old. We had just completed building a custom house when I received my transfer. We hadn't even landscaped the yard. I immediately put the house on the market, and it sold within a week.

A year earlier, Linda and I purchased a twenty-five-foot travel trailer. When we built our house, we ordered a wider than a normal piece of property so that we could set up a spot alongside the house for our travel trailer. We had a concrete slab poured, installed an electrical outlet and piped into a sewer trap so that the trailer could double as a guest house. It worked out well, but we weren't in the house long enough to have any guests.

When we moved to Los Angeles, we hooked the trailer up to our Chevrolet Blazer and headed west from Oklahoma City. The first night we stopped in Flagstaff, Arizona. The temperature was below freezing, so we left the water hose trickle all night to keep it from freezing. As we were leaving the camping area, we had to wait for a herd of Elk crossing the highway. They were magnificent. We continued to Newberry Park, California where we had purchased a house while we were on a short temporary house hunting trip.

Newberry Park was where my brother George (Buddy) lived, and I wanted to live near him, so the kids didn't have to go to a school in Los Angeles. We wanted a rural atmosphere. Our temporary quarters were set up in an apartment two blocks from my brother's house.

The Los Angeles Regional Office was located four miles from the Los Angeles International Airport in Hawthorne, California. In Oklahoma City, I drove a 1974 Corvette but sold it to Linda's cousin because I knew I was going to commute and my Corvette would cost too much to drive. I bought a VW Beetle instead. The house we bought was two blocks from my brother's house. He and a couple of his friends worked next to the Hawthorne

Airport, and we were close enough to the FAA Regional Office so that we could commute together.

The distance from Newbury Park and the FAA Regional Office was exactly fifty miles. Commuting time was one hour going in to work and two to two and a half hours coming home. The traffic was awful. My normal working hours were from six am to two thirty pm in an attempt to miss the rush hour traffic. That only worked going in to work, but not going home.

Vince Bogda was the husband of my brother's oldest daughter, Judy. Vince bought an airplane called a Piper Tomahawk and asked if I could teach him to fly. I agreed so, on weekends we would meet at the Van Nuys Airport where he parked his airplane, and I taught him how to fly. Since he was a part of my immediate family, I asked the FAA if they considered this a conflict of interest and they said no. He was an excellent student and received his private pilot certificate in a minimum amount of hours. He used his airplane to commute from Van Nuys Airport to Hughes Tool and Die Company at the Fullerton Airport. Vince received his Master's Degree from Massachusetts Institute of Technology in Boston (MIT) in Electrical Engineering. He had tested genius. I enjoyed being around him.

For teaching him to fly, Vince bought me a Remington 7mm Magnum Rifle.

My position at the FAA Western Regional Office was as an Operations Specialist (Airplanes and Helicopters). The entire FAA was reorganizing, and I was appointed to head the reorganization of combining the Pacific Region, into the Western Region changing the name to the Wester-Pacific Region. I was required to travel to Honolulu, Hawaii many times to coordinate the merger. Of course, I hated the travel to Honolulu.

My manager, Clyde DeHart was also my manager at the FAA Aeronautical Center in Oklahoma City, and I considered him as the best manager I have ever known. During a conversation, he recommended I buy a computer and learn how to operate it. He said this is going to be FAAs future for conducting business.

The following week I sold a couple of my guns to get the money to buy me a computer. I purchased an Ohio Scientific computer and spent many hours at home learning how to program. There wasn't much software on the market for this computer, so I had to learn to programme.

Learning to program the computer; I would suddenly realize the sun had come up and a new day had begun. Linda hated the computer because it took up so much of my time. I couldn't do it at work because I certainly didn't have the time there.

Vince wanted me to teach him how to fly instruments, which would allow him to fly in the clouds between his home airport, Van Nuys and Fullerton Airport. So, again on weekends, we would fly his Piper Tomahawk around the Los Angeles area practicing instrument approaches into various airports. After flying forty hours of instruments, he took a check ride with a pilot examiner and became a rated instrument pilot.

My brother George (Buddy), Vince and I would go shooting in the desert often. For teaching Vince how to fly he bought me a Remington 7 millimeter Magnum Rifle. When I taught him instruments, he bought me a variable power rifle scope. We saved money by reloading pistol and rifle cartridges along with shotgun shells.

My home was burglarized one day because both Linda and I worked and the girls were in school. My computer was one of the items taken, and Linda didn't complain about that loss at all. In fact, she was ecstatic about it. I replace it though with an Apple II Computer. Again, I spent many hours learning to operate this computer and again Linda complained.

At the FAA's Western Pacific Regional Office, I was assigned to participate on a team to evaluate different models of computers to determine which would be the best for each district office. I traveled to the FAA headquarters in Washington DC where we set up the evaluation team and made arrangements with all the bidding companies for demonstrations. One company was Burroughs.

The Smithsonian Institute Museum of Aviation was across the street from the FAA Headquarters, and every time I was in Washington, I would go to the museum and look for that picture I took. I never found it, even asking for help I never found it.

The FAA evaluation team traveled to Washington, D.C. to attend a conference at Burroughs headquarters where they planned to demonstrate and show their computers. On hand to demonstrate the Burroughs hardware was company personnel specifically trained for this sort of sales pitch. After demonstrating their word processing software, a Burroughs staffer started demonstrating their spreadsheet program. This software looked familiar to me because I had a spreadsheet program on my Apple Computer called VisiCalc and this software was very similar to what they were demonstrating.

To affirm in my mind that it was the same as Visicalc, I asked the instructor demonstrating the software if he could split the screen and set up calculations on both sides. He said no, this software couldn't split the screen. He continued, and I still had the feeling this was what I had on my computer so, I asked him again if the screen could be split and do calculations on both sides and again he said, "No." It was obvious he was a little perturbed with me because he was being interrupted by a novice. Then, attempting to belittle me, he asked if I wanted to come up and try the computer myself. Logical thinking of what two keys would split the screen, I accepted his challenge and went to the computer and pressed the function key along with the W (for windows). The screen all of a sudden split in two and the spreadsheet appeared on both sides of the screen. Everyone in the room went crazy, and I became FAAs hero. The FAA purchased the Burroughs, and I was glad because it was the only computer that contained a hard drive.

Back home my brother and I were driving to Santa Barbara, and he told me things went crazy at Burroughs when someone from the government had disrupted a demonstration by outdoing the Burroughs representative. I laughed and told him; it was me. I then told him the entire story, and he we both laughed historically. I was famous.

Again the Burroughs was the best buy because all the rest of the computers operated using an eight-inch floppy disk. I won a cash award for my participation in the FAAs purchase. Maybe I also got a little recognition for showing up the Burroughs's demonstrator.

Burroughs provided training for everyone at the regional office, including me. I think I pestered the instructor too much because I knew what she was going to teach before she even began. If it were not that I used my sense of humor with her, I believe she would have hated me, but on the contrary, we got along fabulously.

My next big assignment was writing the results of the Division Manager's Performance Evaluation. I felt overwhelmed me until I got the idea of how to use the actual performance standards to show how he exceeded each one. He received an Outstanding Performance Award based on my report.

The district offices of the FAA received the computers and began their training. They were in use for about four months when I received the assignment of being the lead focal point for the Western Pacific Regional implementation the Aviation Safety Analysis System (ASAS). I had to travel to numerous FAA District Offices throughout the US interviewing people for the development of databases needed in a district office. I was assigned an FAA Airworthiness Inspector, Travis Boren and two paid consultants who held PhDs in business administration.

After gathering information from the District Offices, we spend two weeks in a conference room at FAA headquarters in Washington, DC, compiling the data into useful databases. I again received a cash award for my participation. While in Washington, I was asked to put on a presentation to all FAA Regional Directors (number one executive in each of the FAAs regions) explaining the concept of ASAS. I did this cold without having any notes or preparation. I received many compliments, and my efforts were a total success. Whew!

The next time a presentation was required, I gave that assignment to Travis Boren, my Airworthiness Inspector cohort. His presentation was much better than mine and so anytime a presentation we needed, Travis was the one. He was great.

Vince bought a new airplane. It was a Mooney 231 which had a Turbocharged Lycoming Engine. It seated four and had a top speed of one hundred and ninety miles per hour. For insurance purposes, I had to give Vince ten hours of instruction because he did not have any experience with a retractable landing gear aircraft. We conducted this training at the Van Nuys Airport near where he lived. Again, he did an outstanding job, and I signed his logbook showing he received this instruction and the insurance company was happy.

TRAGEDY STRUCK

I was invited by Vince to go fishing on one weekend with him and my brother to Lake Powell in Arizona, but I declined because I promised Linda and the kids to take them somewhere.

On Sunday afternoon I received a call from my niece Judy, asking if I could call someone to find out why they weren't home yet. I placed a call to the Regional Duty Officer who told me a Vince Bogda had just crashed at the Van Nuys Airport and there were two fatalities. I blurted out over the phone, "That's my brother."

I drove to Judy's house and told her all of what I knew. Judy was eight months pregnant with Michael. We then went to my mother's apartment and told her the bad news. We remained with my mother for some time

Conspiracy

consoling her, and then we went over to my brother's house and broke the news to my brother's wife Kitty (Catherine) and his son and daughter George and Joyce. Since my brother's house was only two blocks from mine, I went home afterward.

The following day when I went to work, I went up to the duty officer and apologized for telling him "That's my brother" over the phone but he understood and didn't have a problem with it.

I went home early and stayed there the rest of the week. Someone complained that I interfered with the investigation and the Regional Director, Mac McClure (number one man in the region) called me personally and told me about this complaint and he said not to worry, he straightened them out by telling them I wouldn't stoop to that level. I thanked him.

Two days later my brother's son George came to the house and started asking questions. I took him to the Van Nuys airport to look around. With the remnants of the accident removed, and the entire area cleaned, I went to a house across the street and showed my FAA Investigators identity to the man who answered the door. He was home when the accident occurred and explained what he saw. Not wanting to hear anything further, I took George home.

I obtained a copy of the Air Traffic Accident Report from a friend in Air Traffic Division at the Regional Office. Reading it, I realized air traffic was attempting to cover up the fact the Air Traffic Controller screwed up by giving Vince erroneous and incorrect instructions. The maneuver the controller demanded of Vince caused the aircraft to stall and crash because it was too difficult for any pilot to accomplish without encountering catastrophic results. While giving Vince the instruction in his new airplane, I had warned Vince that if he didn't like a particular instruction from an Air Traffic Controller he had every right to refuse it. However, Vince and as most other pilots always tried to give the Air Traffic Controller what they wanted because it would help them get along. However, this time it killed both Vince and my brother.

Raoul Magana was a very close friend and was a successful attorney with offices that took up an entire floor on one of the buildings in Century City, California. Magana, Cathcart was the name of the firm. Cathcart is the adopted son of Raoul. Raoul and I first met when he signed up with me to learn how to fly. We flew together for only about nine times and for some unknown reason he stopped. Raoul's reputation as an attorney was impeccable throughout the entire US and specialized in medical malpractice cases. His everyday vocabulary was way over my head.

I recommended Judy to see Raoul and to tell him about me being her uncle. She did, and Raoul agreed to take the case. His associate attorney represented Judy in the court against the FAA and asked that I testify on behalf of my niece, Judy. I researched and found I was allowed to testify because Judy was my immediate family. Her attorney issued me a subpoena forcing me under the law to testify during the trial.

I received telephone calls from various U.S. Attorneys in Washington, DC threatening me if I didn't stop my niece there was going to be trouble. Naturally, I couldn't stop my niece because I had no control over her and I had nothing to do with pursuing the case. I was only asked to testify on behalf of Vince because I taught him how to fly.

Eventually, my niece won the lawsuit, and as the judge called it, one hundred percent fault was on the FAA, and she was awarded one million dollars for Vince's death. However, this judgment didn't become final for a year and a half after the trial ended.

I received a telephone call from the Division Manager of Southwest Region asking if I would like to move there and become the FAAs, Helicopter Specialist. I was so tired of that daily commute in the Los Angeles area I accepted and moved the family to Grapevine, Texas. Grapevine is a quaint little town located north of and almost equidistance between Fort Worth and Dallas. My office was in a WWII helium factory near the Fort Worth airport. The building was almost falling and desperately in need of repairs. I found termite traces on the windowsill next to my desk. It was very depressing.

The division manager was Roger Knight, and we had known each other before I came to the South West Region. He asked me if I could write a computer program linking all the district offices together by computer so that all of them could track their legal cases to determine, their stage of completion. Further, he wanted me to monitor all the district offices receiving the Burroughs computers and fix any operational problems. I enjoyed this assignment because it gave me full use of the regions Beechcraft Kingair C90, a twin-engine turboprop airplane. I traveled in the Kingair to every district office when they encounter a problem with their computers.

A report came out from Washington Headquarters showing computer downtime for each region. Percentage-wise, our region had the lowest by far at 3 percent. The rest of the regions ranged from 50 to 70 percent downtime.

Conspiracy

I finished linking the district offices with the Regional Office's Legal Department. The FAA has never been able to track the enforcement cases to completion. Eventually, the FAA headquarters in Washington, D.C. sent a team to evaluate my program. After their review, they adopted my ideas for a national program. I received another cash award for me creating this program.

Roger asked that a Flight Test Engineer Pilot accompanied me when I rented a helicopter to fly and visit the Oil Rigs in the Gulf of Mexico. Our purpose was to certify the airborne radar installed in the helicopter to avoid obstructions. Roger wanted approval to fly a helicopter in the clouds while searching for the destination Oil Rig. Except for Roger, we all felt this was dangerous, but Roger felt if he could get approval it would make him look good in the eyes of higher management in Washington.

Jim Arnold, the Flight Test Engineer, and I traveled to New Iberia, Louisiana where we rented a twin-engine Sikorski S-76 Helicopter. Jim, not being qualified in helicopters, sat in the back. So I flew it along with the company's pilot. We left the airport and leveled off at an altitude just above the trees. I increased our speed as fast as the helicopter would allow. We went skimming just above the trees playing like we were a gunship. We looked down at the water and saw many alligators just floating, waiting for something to go by so they could eat. There were many small empty huts throughout the area that people use for fishing and poaching alligators.

Once we got out into the Gulf of Mexico, the other rated pilot took over control allowing Jim to ride in one of the pilot seats, allowing him to test the radar and evaluate its ability to detect metal objects in the water. I was in the back playing with the Loran Navigation Unit.

The Loran was a low-frequency radio receiver that received signals from three separate transmitters located somewhere in the area. The Loran would then triangulate the signals and provide the pilot with the ability to determine his/her exact location accurately. The Loran was the first radio to accomplish this. Today the Loran has evolved into the GPS which uses Satellite Transmitters instead of the low-frequency units of the Loran.

Watching the operation from the rear of the helicopter made me realize we could use the Loran Navigation Receiver as primary for course guidance while utilizing the radar for obstruction avoidance. All while executing an instrument approach procedure so that no matter which way the wind was blowing, the helicopter could make an instrument approach safely to an oil rig in the Gulf. This new procedure would save hundreds of thousands of dollars for both the FAA and helicopter operators throughout the world. They would benefit by being able to fly in the clouds by instruments and be able to navigate to any oil rig safely and always into the wind.

In the past, operators had to spend thousands of dollars getting the FAA to fly each approach into an oil rig they wanted approval and in only one direction. With this new procedure, the operator only had to demonstrate their knowledge and skill flying into one rig because it was generic to all the other rigs.

We got back to Fort Worth and had a meeting with Branch Managers and Roger and laid out our findings. Roger was ecstatic because this satisfied his ambitions to show higher management in the FAA his new concept.

After a year and a half in Ft Worth, I was ready to leave because I didn't like Texas or Ft. Worth. I had applied for a supervisor position at the Scottsdale, Arizona Flight Standards District Office and was selected. Roger was on vacation and didn't know I had been selected and was leaving. I packed up and left Ft Worth. I got word later he wouldn't have let me go if he had not been on vacation. It was too late, Linda and the kids and I were on our way to Scottsdale, Arizona.

SCOTTSDALE, ARIZONA

We arrived and stayed at the Pointe Resort at Tapatio, Arizona. This resort was unbelievably beautiful, and the kids loved it. For some reason, Linda was backward in her directions, and when she felt she was going north, she was going south. It took her a couple of weeks to overcome her problem.

She was busy house hunting and found a house with a pool only one mile from Scottsdale Airport Office.

Eldon Gubler was the manager of the Scottsdale Flight Standards District Office (FSDO). Another new employee assigned to the clerical team arrived by the name of Chris. Everyone was in a meeting, and Chris didn't hesitate one second before jumping in answering phones and picking up the responsibility of running the office while everyone was unavailable.

Chris and I talked later about working with Eldon for getting the FAA Office of the Year Award before he retired. As we worked out our plan, we drafted other reliable people who helped tremendously. Before the year was over, the Scottsdale FSDO had won the National Award for being the "Outstanding District Office of the Year Award."

The office took delivery of many computers for secretarial purposes, but no one knew how to operate them except me. The office stored these new computers in boxes until the FAA found it convenient to teach everyone how to use them. So, I asked Eldon if I could use the office on my own time at night to teach everyone how to use these new computers.

The computers were PC clones. A friend Bob, who lived in Burbank, California, showed me how to build and test clone computers. So, I bought the components and built me a PC clone. I sold the Apple Computer and Bob, and I started a small business building and selling computers, so we knew them very well.

Eldon agreed so, after work when the district office was closed, we set these computers up in the conference room, and I taught everyone how to use a word processor and the tasks of maintaining disks. We did this every night for a couple of weeks. In the end, our secretarial personnel became very proficient in the use of computers.

FAA'S CONSPIRACY BEGINS

The court in charge of my niece's case finally reached a final decision. She won. The decision was that the FAA was at fault one hundred percent and Judy awarded one million dollars. As soon as this settlement became known, the officials in the FAA came after me like a rabid dog, blaming me for everything.

I made friends with a local pilot Ray Morgan that flew a corporate turboprop airplane. The name of the company Ray flew for was Life Shares, and Jim Fail was the CEO. They sold the Merlin and bought a Sabreliner. Ray, knowing that I taught in this airplane at the FAAs aeronautical center asked if it would be alright if I accompanied him on some trips teaching him everything about the airplane. I checked with the Regional Office's Legal Department in Los Angeles, and they told me there was a letter from FAA Legal Department in Washington DC that stated, so long as I didn't accept any remuneration of any kind it was legal and OK. The Region sent me a copy. We went on a couple of trips together, and Ray did an outstanding job flying the aircraft.

One day Ray became sick and asked if I could take his place on a couple of flights until he recovered. I said OK and took annual leave as I did on all the other flights as well, so I wouldn't get accused of defrauding the government of annual leave.

We flew back to Washington, DC and my co-pilot was a retired TWA Chief Pilot. We flew from Dulles, Washington, DC to Boston, and back to Washington and then to Scottsdale. When we reached cruising altitude, Jim called me to the back of the airplane and told me he was expected to sign some papers and asked if I could use the airplanes telephone to contact someone and find out what was going on. I did and found that everyone was waiting for him at Love Field in Dallas, Texas. At the time we were over Indianapolis, and Jim asked if we could divert to Dallas, and I said yes. We landed about an hour later, and people were there waiting for him to arrive. I told Jim I would have the airplane all fueled and ready to return to Scottsdale when he got back.

Jim returned in two hours, and we departed Love Field for Scottsdale. Once we were at cruising altitude, Jim called me to the back of the airplane, and he told me that diversion to Dallas made him two point four billion dollars. I was shocked, but Jim deserved this kind of success. He was a hard worker and a great individual.

When we moved to Scottsdale, Arizona, Linda wanted to work as a Flight Attendant for America West Airlines instead of what she had been doing working as a loan officer in a bank. I felt she had to do what she wanted and I told her I would help in any way possible. She graduated from training and started flying. The girls were old enough they could take care of themselves. It wasn't long after she started flying that she told me she wanted a divorce not giving any viable reason. I said ok, and her and the two girls left. I would take the girls in with me whenever they wanted. In ten years of marriage, I became to love both Stephanie and Melanie dearly and for them to leave, broke my heart. It didn't' bother me at all Linda leaving.

I wrote another computer program designed that electronically stored every responsibility within a district office. With this information, an inspector can plan and schedule their annual surveillance program and calculate how many hours it would take. When I got it up and running, I was asked to come to the regional office and brief everyone on the program and how FSDOs could use it.

I was in the regional office and the Assistant Division Manager Dave told me he knew of the letter allowing me to fly the Sabreliner but he didn't like it. I never uttered a peep because they could turn anything I said into insubordination, and I didn't want to give them any ammunition. It became obvious they were after me because of my niece winning her case against the FAA.

Whenever someone needed a Sabre specialist, however, the region would send me because I was the only Sabre specialist current on the airplane. When this assignment was complete, they would again start threatening me about flying the Sabreliner again. Finally, Larry, the office manager came to me and asked that I not fly that airplane any longer because the region is making his life miserable. For him only, I stopped flying.

Other district offices in Iowa, Colorado and New Mexico heard about my computer program. They contacted my regional office and asked if they could borrow me to install this program on their computer system and that they would provide the necessary travel expenses.

The Division Manager in our region heard I was building computers and selling them. They accused me of a conflict of interest and initiated an investigation against me. This investigation lasted two years, and they found nothing. The regional office became frustrated with me and sent me orders transferring me to the regional office. I felt if I went to the regional office they would fabricate charges against me and terminate my employment. I did not want to go back to the regional office and to that commute. I had ninety days in which to report or be terminated.

WORKING IN THE ORIENT

The following week I was contacted by the regional office assigning me to inspect the Flight Standards National Flight Office in Tokyo, Japan. Their reasoning was, I was current in the Sabreliner which is the same airplane I flew for my friend and was also used by Flight Inspection personnel. This is a classic example of hypocrisy at its finest. The regional management kept badgering me to quit flying my friend's corporate Sabreliner, but the FAA kept taking advantage of my currency in the aircraft by assigning me to inspect a Flight Inspection Facility, in their Sabreliner.

The regional office wanted me to travel to Tokyo, Japan without paying the airline. To do this, I was to conduct an Enroute Inspection by riding in the cockpit observing the flight crew perform their duties. The flight from Los Angeles to Tokyo by Northwest Airlines would take fourteen hours. I didn't want to sit in the cockpit for fourteen hours, so I refused. They agreed and provided me with the money to buy a coach ticket.

I rode America West Airlines to Los Angeles and then on Northwest Airlines Flight One. To take advantage to the situation, I used my FAA credentials to ride in the cockpit with the flight crew. The flight would take approximately fourteen hours, so when we reached our cruising altitude, I informed the flight crew that they didn't want me up there, so I asked if I could get a seat in the back. They agreed and made a call to the Lead Flight Attendant and asked to set me up with a seat in first class. I lived well.

I notified my friend Greg Knapp who had been living in Tokyo for the past nineteen years that I was arriving and I would be there for a while. He answered me and said yes come ahead and to plan on staying with him.

The airplane arrived in Narita Airport, Tokyo, Japan after fourteen hours of flying. Fortunately, I slept in a crew bunk with paper sheets, so I was alert when we landed. Greg was there to meet me, and we headed for Tokyo by bus. Construction of the new train system was not complete by the time I traveled from Tokyo to Narita Airport.

Greg was working for a movie production company called Telledan out of Burbank, California and they paid the rent for his condominium. It was huge especially in Japanese standards and very comfortable. His wife Marsha was home when we arrived, and we had a small snack. They had adopted a Japanese baby boy who had just finished his nap. He was really cute, and his name is Holden.

The next day Greg took me to the Kodokan, Judo's international headquarters. Tokyo is where I wanted to come when I finished Naval Intelligence School but was sent to Turkey instead. When Greg and I were practicing at Seinan Judo Dojo, the world champion was a man named Diago. Greg introduced me to him at the Kodokan, and I was amazed he still looked mean, but his demeanor was one of a perfect gentleman. I was honored meeting him.

Since we were kids, our Japanese friends in Judo told us about the revenge of the Forty Seven Ronin (samurai) also known as the Akō vendetta. In the 18th century, a group of samurai warriors was left leaderless, when their leader Asano Naganori was compelled to commit ritual suicide (hara-kiri) for assaulting a court official. The ronin avenged their master's honor by killing Kira, after waiting and planning for a year. In turn, the ronin were then themselves obliged to commit ritual suicide for committing the crime of murder. In Tokyo, there is a memorial honoring each of these ronin. Their bodies cremated, and their ashes placed in a vase, now located in this memorial. One of the ronin was only fourteen years old.

FIGURE 21 - SENGOKU-JI

Family members of the ronin placed incense next to the inscription identifying each member, and there was incense burning when I visited this place of honor.

The next day I reported to the Flight Inspection District Office (FIDO) in Tokyo. I briefed all the flight personnel on my mission. The aircraft we would be using is a Sabreliner 40 Model, and we should not have to deviate from their normal mission just to please me. I assured them I was there to help improve their operations. I knew two of the people I would be flying with from having them in my recurrent classes in the Sabreliner in Oklahoma City.

We drove to where the aircraft parked at the Haneda Airport. After preflight and filing of the flight plan we departed and headed for the Airforce Base in eastern Korea. There were four of us on board and after landing we checked into a local hotel where we spent the night. In the morning we departed and flew up north to within a few miles of the 38th parallel. I'm sure we were on the North Korean's radar. I wanted to tell the Air Traffic Controller to make sure we didn't get too close, but I didn't.

At this air base, we watched the U2 spy planes taxiing with armed guards walking alongside until they were inside a hangar.

We spent the entire week in Korea checking navigational facilities and then returned to Haneda Airport in Tokyo.

To get back to Greg's place, I had to take a train, and it was called Inokasura Line. The train station did not have any signs with English characters. When I was active in Judo I was learning to speak, read and write Japanese. I thought back to my lessons, and I saw two Japanese characters I recognized. Together they made the word "Ino." I took a gamble and boarded the train, and it took me to Greg's house.

The following Monday we departed for the military base in Okinawa. There we saw a Douglas F4 Phantom with four red stars on its fuselage. We refueled, filed a new flight plan and ate lunch. We then departed for Clark Air Force Base in the Philippines.

We were navigating in the clouds on an Omega Long Range Navigation system. On the front of the control panel was a placard which read, "VFR ONLY" meaning we were not supposed to be flying in the clouds (IFR) using this navigation unit. Legally, we were only supposed to use this means of navigating when we were clear of all clouds. I noted this and planned to write this up in my report. In essence, this flight was violating the Federal Aviation Regulations, but I didn't have the authority to stop the flight.

There was also a regulation that all FAA aircraft must have dual navigation units (VLF Omega) installed to fly in the clouds. This aircraft only had one, and this one wasn't allowed to fly in the clouds. The FAAs mentality was; regulations are for others and not the FAA.

We landed at Clark Airforce Base in the Philippines where we refueled and had lunch. We boarded a Philippine national as an observer. He knew the entire area and showed us the famous landmarks of WWII. He showed us the route General MacArthur took to escape the Japanese invasion, Corregidor, the Baton Death March, etc. The Philippine government tossed millions of silver dollars into Manila's Bay to keep the Japanese from getting this money.

Jacque Cousteau did a TV special on retrieving the silver dollars from Manila Bay.

We remained at Clark Airbase for a week staying in a hotel that had ten-foot walls surrounding the hotel. There were two guard towers located at two of the corners of the walls. Guards manned in these towers at night for protection of the hotel's guests from whatever.

Just outside the gates of Clark was a bar and strip club owned by a retired Master Sergeant. The club had no morals whatsoever. One of the dance girls, I was told, was only twelve years old. The others were all under the age of sixteen. When we saw this, we all left and went back to the hotel.

The taxicabs in use were jeeps that were made in the U.S. and left over from WWII. The Philippines modified them by extending the rear body to carry additional people in the back and painted them in different and bright colors. These paint schemes had no pattern at all, just paint that didn't blend. They looked horrible.

After leaving the Philippines, we circumnavigated a large typhoon using the VLF Omega which was designated not to be used inside clouds. Again, I planned on putting this in my report.

The flight took us to our next destination, Guam which is a US possession. There we again flight testing the navigational facilities on the island. It only took us one day, and we were then on our way back to Tokyo.

I said my fair wells to Greg and his wife Marsha, and I left Narita Airport, bound for Los Angeles and then home in Phoenix.

In the FAA office, the investigator assigned to find a conflict of interest was still diligently searching for evidence but was not having too much luck. When the investigator found that I had used government facilities to teach the secretaries at the FAA FSDO, he thought he had found something to prosecute me on for misuse of government equipment. However, when he interviewed the office manager, he found I had permission to teach FAA personnel in the exam room on my own time after office hours.

For the past two years, the complexity of my position as an Aviation Safety Inspector Supervisor in Operations had increased from a GS-14 to the level of GS-15 pay grade. However, the Division Manager, Bill Williams refused to increase my pay grade to the level commensurate with the position. My counterpart in Maintenance had been a GS-15 for five years because his position ranked at the GS-15 level. So for two years, I was underpaid, and I can only assume this was because of the FAA's vendetta against me.

Meanwhile, I checked with the Human Resource Manager at the Regional Office and asked if the transfer letter moving me to Los Angeles against my will was still valid. She said it was. Then she said I was eligible to retire if I wanted even though I was not at the retirement age. I found

I would lose a substantial amount of retirement money if I retired early. However, I felt I had no choice, so I retired. I feel the letter I received was just a means to fire me if I didn't move to Los Angeles.

I had heard Eastern Airlines was hiring and with my jet airplane experience, maybe I could get a job flying with them. I contacted Eastern Airlines in Atlanta, Georgia and told them of my background and experience. The sent me an airline ticket and made a reservation for me on Eastern Airlines leaving Los Angeles. The following day I was on my way to Atlanta to be interviewed.

According to the letter I received transferring me to Los Angeles; I didn't have to report for another three weeks. I landed in Atlanta and went immediately to the employment office of Eastern Airlines. I had an interview with a retired captain who volunteered to conduct employment interviews. I told him of my experience as an FAA Supervisory Aviation Safety Inspector and as a flight instructor at FAA Aeronautical Center in Oklahoma City. When the interview was over, the captain told me he was going to recommend me for the position as captain of a Boeing 727. I was a little nervous because being a captain of a Boeing 727 had enormous responsibilities and it intimidated me.

We established a reporting date for me, and I asked if the airline would allow me to remain with the FAA and I would take annual leave enabling me to go through their training. Their answer was "yes." They didn't mind at all because Eastern Airlines had temporarily terminated flights in and out of Phoenix, Arizona, and the FAA wouldn't be able to accuse me of having a conflict of interest.

I returned to Phoenix and called the Human Resource Division Manager and asked for a meeting. Unbeknownst to anyone except my office manager, I flew to Los Angeles and met with the HR Manager. I asked if I had enough years with the FAA to start receiving an annuity if I retired under these circumstances of forcibly transferring me to the Los Angeles Regional Office. She said, yes I could retire. However, the amount of my retirement (money) would be affected because my age and years of service did not add up to eighty.

I signed the necessary paperwork and retired. After everything was complete, I got word that the division manager had rescinded the transfer letter attempting to remove the justification I needed to retire. However, the paperwork was accepted before he rescinded the letter, so my retirement remained in effect.

Conspiracy

What bothered me most was that I was a dedicated employee. Many times I donated my time and computer knowledge to the FAA for nothing. I further donated my computer programs I wrote to make the FAA function better. In the process, I received numerous awards for the work I did. I feel betrayed. Three of my programs remain throughout the FAA. The enforcement program: the PTRS program, and the databases that were a part of the PTRS.

Throughout my nineteen years of service with the FAA I received the following:

Cash Awards	Letters of Commendation	Letters of Appreciation	Certificate of Merit	Special Achievement Awards
4	2	13	1	7

Chapter 7

LEAVING THE FAA

Miami, Florida was the training base for Eastern Airlines. They got us rooms at the Dural Country Club and golf resort. There were six golf courses at the Dural Country Club, but the humidity was horrible, so golf was not the usual past time. The food was outstanding, and there was always plenty to eat.

The training lasted five weeks, one week of indoctrination, three weeks of B-727 systems and one week of aircraft simulator and the check ride or flight test.

As the days went by, Eastern began rehiring individuals with captain experience. I felt relieved when they removed me from the captain's training list. I didn't mind this at all, in fact, I welcomed it.

We passed all the tests from Indoctrination, systems ground school and the simulator when I found out La Guardia Airport in New York City was going to be my base. La Guardia was the armpit of New York, and being raised in New York, I knew I wasn't going to like it. I was not looking forward to this assignment, but I figured to make the most of it.

Not two months in New York, Eastern asked for volunteers to switch to the DC-9. The base for the DC-9 crews is in Atlanta, George. My mind was up before he finished talking. Then back to Miami for the DC-9 transition. Three weeks later I went to St. Louis for simulator training at Flight Safety.

The check pilot was a retired DC-9 captain from Eastern Airlines that reminded me of the Bulldog the U.S. Marines Corps had for a mascot. He was a real gentleman though and very congenial throughout the training, but he sure looked tough though.

In the simulator, I was in the co-pilot seat executing an instrument approach with one engine shut down, and he came walking in. I had everything perfect while on the approach, and he complimented me regarding my abilities. From then on he could do no wrong. I passed the training without a problem and reported to Atlanta for co-pilot duties.

The commute for me from Phoenix was much easier and shorter than trying to commute to New York City. Since I was not very senior, my bidding lines were that of a reserve pilot, meaning I could go anywhere and fly with any captain. I liked flying reserve because it kept me from getting bored flying the same destination with the same crew for thirty days. Flying reserve gave me a better experience than the mundane trips a scheduled line had. They were boring because they were the same, day in and day out.

The one thing I didn't miss was the FAA's vendetta they had against me. I was always waiting for them to try something against me while with Eastern, but they didn't so far as I knew.

When I was with the FAA, I would ride in the cockpit watching the flight crew performing their operation. I always thought this would be great if I could be an Airline Captain. The one thing the Eastern pilots did on a regular basis was to fly the airplane themselves from takeoff to altitude and then engage the autopilot. On the descent, they would hand fly the airplane, execute the approach and land. I was impressed. Every once in a while the Captain would use the autopilot if the weather was bad giving them the advantage.

I would always manually fly the airplane when leaving a cruise altitude, execute the approach no matter what the weather was and land the airplane. My respect for Eastern pilots grew every day I was flying with them.

On my days off I would travel back to Phoenix to spend my time at home. During these days off I met and dated a woman named Nancy. After about six months of dating, we went to Los Vegas and were married. She had two sons Billy and David, both grown and out of the house. Nancy had a granddaughter Ashley. She was eight years old and was the daughter of Billy. She would stay with us a lot. What a sweet girl. We got along great, and I loved having her with us.

Having a reserve schedule gave me flexibility so that I could take ten days off in a row and spend that time in Phoenix. Dick Thomas operations unit supervisor at the FAA FSDO contacted me while I was at home asking me if I would accept a position as a Pilot Examiner for the district office.

A pilot examiner for the Federal Aviation Administration did the same work as an FAA inspector when certifying pilots. An applicant trained as a pilot would present themselves to an examiner for final certification. The examiner would administer an oral examination and if all went well would then fly with the applicant. If the applicant passed the flight portion of the examination, the examiner would issue the applicant a pilot certificate. Pilot examiners receive money from the applicant for this examination.

I accepted his offer, and I contacted Eastern Airlines to see if they would allow me time off to attend an examiner training slot in Oklahoma City. They agreed, and I trained as a pilot examiner. Dick Thomas also wanted me to administer ratings to pilots seeking a rating in a Sabreliner.

I satisfactorily completed all the training and was certified as a pilot examiner, including the Sabreliner. I went back to Eastern Airlines and continued flying with them as a co-pilot.

In January 1991, Eastern Airlines declared bankruptcy and shut down. I went back to Phoenix and administered pilot certification to applicants at the Scottsdale Airport.

RETURNING HOME FROM EASTERN

One of the requirements for a pilot examiner to conduct certification checks in a Jet Airplane is an FAA inspector must observe giving a check to a pilot once a year. To rent a Sabreliner would cost around $500 to $600 an hour. To defer this cost, the examiner waits until a pilot who normally flies a Sabreliner for a company is ready for an annual check. The pilot uses his company's Sabreliner for the test. The FAA inspector then rides in the jump seat, and observers the examiner conduct the flight check to the pilot.

My annual check was coming due, and a local pilot asked me to conduct his annual check. I agreed and notified the FSDO of the upcoming check. I contacted the FAA Inspector in charge of my pilot examiner activity Mike Warth and told him of this plan to have the FAA monitor this check. I was told by Mike, that there would be an inspector available on the date and time we arranged.

When that date and time came arrived, I went to the FSDO office and told no one knew anything about this check and Mike Warth was out of the office. I told my applicant of this, and we reschedule it for the following week. Again, when the date and time came up, no one was available to observe this activity. For a third time, this happened, and the individual went elsewhere for his check, and I could no longer give checks in the Sabreliner.

Mike Warth told me, the office manager, came into his office and told him not to schedule this check but do not tell me. Each time we had scheduled the check, the office manager, told Mike not to follow through with my request. I wrote a letter of complaint to FAA Headquarters in Washington DC, The answer I got back was in his exact words, "Awe, he wouldn't do anything like that." Nothing else happened. With this attitude, it was obvious someone was covering for his actions.

I received notice from the FAA District Office Manager, Gary Koch, that he wanted to see me. He told me Tom Accardi was not going to renew my pilot designation.

Tom Accardi is Director of Fight Standards Division in Washington FAA Headquarters. I never met the man, so I asked why Accardi is coming after me. All he told me was that I had a conflict of interest when I was working for the FAA. I denied this and said an investigator from the security division from the regional office worked on this for two years and couldn't find anything.

I contacted Senator John McCain and informed his staff member that the FAA did not have any justification not to renew my pilot examiner designation. The staff member agreed with me and my concerns, however, Senator McCain did nothing and allowed this to happen without even voicing his opinion. When my renewal came due, I hired an attorney, Mr. Tom Toone to represent me in my hearing at FAA Regional Headquarters. He and I traveled to the regional office in Los Angeles to fight these allegations. Present was the division manager, the regional attorney, Tom Toone and myself. The division manager accused me of having a conflict of interest when the FAA employed me, and this justified not renewing my pilot examiner. The Division Manager did not present any evidence of the so-called conflict of interest other than them just sitting there and verbally accusing me. My attorney Tom told me later that he realized this was a classic example of a kangaroo court and there wasn't anything we could do about it.

Tom wrote his report and formatted it to present the allegations to the Ninth Circuit Court of Appeals in San Francisco. After waiting two months they determined they weren't going to decide this matter but did make the statement that it was obvious Tom Accardi, AFS-1 had a personal grudge against me. As a pilot examiner for the past year, the FAA FSDO never received one complaint stemming from an applicant. It was obvious it was the FAA's vendetta resuming after my being out of the area for two years flying for Eastern Airlines.

MUTO AVIATION

I received word that a company in Glendale called Muto Helicopters was looking for a helicopter instructor who could certify the company under Federal Aviation Regulations, Part 141 as an Approved Flight School. I had certified other companies under this regulation, and it was a fairly simple process. I approached Mr. Muto with my resume, and he hired me with a salary of one hundred thousand dollars a year.

To welcome me to his organization, Mr. Muto, along with three other employees and me, flew to Tokyo, Japan to visit his factory. We traveled the famed Bullet Train from Tokyo, north to a small town called Koga. It was spring, and the cherry blossoms were in full bloom. The beauty of it all was breathtaking. The cherry blossoms were a part of a Japanese Temple Grounds. Inside the walls was a statue of Buddha, and two wooden temples with bells that had heavy logs suspended by two ropes designed to ring the bells. The grounds were impeccable, and everything was beautiful.

The week after coming home, Muto asked me to fly the office manager Mitsuko to demonstrate the helicopter to the Yuma County Sheriff's Office in the hopes of making a sale. We flew the demo Schweitzer Helicopter 300C to Yuma. Our route took us south to Interstate 8 and then due west to Yuma. We demonstrated the helicopter's capabilities to two deputy's, and in two hours we were returning to Phoenix. We flew over a cattle feedlot at one thousand feet above the ground. The smell from the feedlot was overwhelming, and I said something to Mitsuko about the smell, and she started laughing. I asked why she was laughing and she said she thought the smell was coming from me. We both laughed hysterically.

Nancy and I were driving along a three-lane street heading home when a red traffic light turned green, and I went through the intersection passing a pickup truck stopped in the right lane. I heard the truck's engine roar in an attempt to get in front of me because the three lanes merged into one due to a construction area ahead. The traffic light up ahead in front of us had turned red, so I stopped. The truck pulled up behind me and started blowing his horn. I ignored him and when the light turned green; I went through the intersection and left the construction area behind. The road opened up to three lanes. The truck sped around my left cut me off and forced me into a parking lot. We stopped, and both got out of our vehicles, me standing alongside mine and him walking towards me. He was much bigger than me and much younger, I was fifty-seven, and he looked to be in his late twenties.

He came up to me and started yelling, and I just stood there not saying a word. Then he poked a finger into my stomach and said, "You're nothing but a fat old fart." At that point, I just turned to walk away, and Nancy yelled something from inside the car. I turned, and this guy hit me from behind and caught me on my right cheek as I turned. I lost my temper and used judo to throw him to the ground. I should never lose my temper, and to this day I don't know what I did but the next thing I knew I was on top of him punching the hell out of him. After coming to my senses, I stood up and took notice of him as lay bleeding all over the place. I was expecting him to come after me, but instead, he got up and ran away.

I got back in my car and started pulling away when I realized he was taking my license plate number. We arrived home, and about thirty minutes later two Phoenix Police Officers came to my door. I let them in, and one officer talked to Nancy in the living room while the other sat at the dining room table and talked to me. I explained what happened and he said it was almost the same as the other guy with the exception the complainant said I hit him with something. I said, "Yes, I did," my knuckles. When I showed the officer my cheek, and the red mark on it from him hitting me, the officer looked at my cheek and said this is nothing; you should see the other guy he's a mess. He had to go to the hospital for repairs. I felt bad because I should never lose my temper.

Muto paid for a double page ad in a Japanese Aviation Magazine. He had a photographer take my picture and put it in this ad, and my picture took up most of two pages. It took me by surprise, and it was quite impressive.

I took a few days off from work, and I flew my wife Nancy, her sister, and brother-in-law to Portland, Oregon. We rented a car and drove east into the Columbia River Gorge. We turned left on highway 221 and drove to Meade, Washington, where Nancy was born. It was a small town and very friendly. It had been too many years since Nancy had lived there and everything had changed, including the residence.

When I returned, Muto asked for a meeting. While I was up north, the FAA Office Manager, Mr. Gary Koch called and asked Muto if they could meet for a few minutes. In this meeting, Mr. Koch told MR. Muto that so long as I was employed there, the FAA would never certify the company for an approved Flight School Certificate. Muto fired me.

Within a year, Mr. Muto shut the doors to his business and returned to Japan. Just before this meeting with Gary Koch, Mr. Muto had my picture taken, and it appeared in a double page ad advertising his Helicopter Company in Glendale, Arizona. I do not know what this ad cost but it shows Mr. Muto had no plans of firing me until he received the threat by the FAA.

Gary Koch and I used to work together in the Los Angeles Regional Office, and I considered him a friend, but when he transferred to Washington Headquarters, he obviously became part of the Establishment. He did not have the fortitude to stand up to anyone and has always done things behind someone's back. He has not lost his touch.

DOCTOR HAL PRICE-MIRACLE ON 24TH STREET

It was sometime in August of 1991 when I received a call from Dr. Halford (Hal) Price, a General Practitioner in South Central Phoenix, Arizona.

I had just lost my job as Vice President and Director of Operations for Muto Helicopters in Glendale, Arizona, thanks to the Gary Koch, FAA Office Manager at the FAA Flight Standards District Office Scottsdale, Arizona.

FIGURE 22 - NORTH AMERICAN SABRELINER 60 MODEL

Conspiracy

Dr. Price owned a North American Sabreliner 60 Model, N169RF, otherwise known in the military as a T-39. It is a twin-engine jet airplane with a seating capacity of eight passengers and two crewmembers with a top speed of approximately 550 miles per hour. Normal cruise speed is approximately 520 miles per hour.

Dr. Price contacted the Federal Aviation Administration (FAA) in Phoenix, Arizona regarding his suspicion that his airplane had been sabotaged and wanted to know who in the Phoenix Area knew Sabreliners. The person he talked to in the FAA office referred him to me because being retired I used to teach Sabreliners to FAA inspectors at the FAA Aeronautical Academy located in Oklahoma City, Oklahoma.

Dr. Price called and requested I come to Sky Harbor Airport in Phoenix to inspect his aircraft. I arrived, and Dr. Price opened the security gate and guided me to his hangar. He explained that he and his instructor pilot, Mr. John Bumpus were taxiing for takeoff when he realized the nose wheel steering was not functioning. Fortunately, they were able to stop the airplane before it hit anything. The nose wheel steering on the Sabreliner is electrically actuated and hydraulically operated and is controlled by a button on each of the control wheels for the pilot and co-pilot.

The nose wheel has two major components, the strut collar, and nose wheel. The strut collar holds the electrical and hydraulic actuators and linked with the nose wheel by a locking lever. With a tow bar connected, the locking lever rotates out of its notch allowing the nose wheel to rotate completely around 360° for maneuvering.

When employed by the FAA, my position was Aviation Safety Inspector in operations. One of my responsibilities was to investigate aircraft accidents. Taking accurate photographs was a key element during my investigations. I used a Nikon camera with many lenses and attachments for taking close-ups of various components to reveal their condition.

I inspected the nose wheel and found the spring for the locking lever on the opposite side preventing it from engaging in the notch. There was no deformity in the spring which would happen if something forced the spring into its present position. Inspecting the two bolts holding the bracket for the locking lever in place, revealed someone loosened the bolts and repositioned the spring from under the lever. I came to this conclusion because I compared the threads of the two bolts holding the lock lever in place to other bolts in the same area. The other bolts had dirt and dust on them indicating they were untouched. The other bolts had no dirt on the threads indicating someone had removed them.

The previous week, Dr. Price severed relations with a company that was hired to manage and maintain his airplane. They neglected to do mandatory inspections on the Sabreliner and caused the aircraft to be in an un-airworthy condition and illegal to fly. There were only two keys in existence to the hanger; Hal Price had one on his keychain, and the other was in possession of the contracting company.

At my recommendation, Dr. Price hired an FAA-certified mechanic to repair the nose wheel steering and complete the missing airworthiness inspections, bringing the aircraft back up to legal status to fly.

The following week, Dr. Price took the mechanic who was employed by this operator to do maintenance on the Sabreliner to breakfast. During their conversation, Dr. Price told the mechanic about the nose wheel steering and what I had found upon my inspection. The mechanic could hardly speak, didn't finish his breakfast and left the restaurant. The next we heard this mechanic had moved out of the state back east somewhere. Dr. Price never heard anything about him since.

I took photos of each component and enlarged them, wrote an analysis and because the FAA is responsible for all aircraft registrations I sent the report to the FBI to investigate. A short period later, the FBI sent my report back and stated since no one was injured, they were too busy to investigate. Case closed with no action.

Since I was rated and current in the Sabreliner, Dr. Price hired me to fly with him as safety pilot and instructor. We traveled to a small town outside of Cleveland, Ohio to inspect a golf ball factory he was planning to invest.

On November 5th, 1991, we departed Sky Harbor Airport in Phoenix and flew to just outside Cleveland, Ohio. Dr. Price was at the controls, and I was in the co-pilot seat. Six passengers were on board including Dr. Price's pregnant daughter Ginger and her husband, a friend of Dr. Price's, and a business associate with his two sons.

The prevalent winds that day enabled us to fly non-stop to Cleveland, Ohio. We spent two days at the Golf Ball Factory and departed for Phoenix at night. We arrived at the airplane which had frost all over its wings and controls. An airplane with frost on its controls is extremely unsafe to fly because it disturbs the airflow over the wings reducing its ability to generate lift. Also, since the frost and ice is frozen water, it increases the weight of the airplane which again reduces its takeoff performance.

The airport had limited facilities and had no means of removing snow or ice with approved de-icing fluid. I had no other choice but to take a water hose and spray all the frost and ice off the airplane. This procedure is the safest method at our disposal. So long as we didn't delay our departure and were airborne before the water had a chance to freeze again, it would be safe. It takes a while for the water to freeze because the warmer water from the hose not only thaws and removes the frost and ice but it increases the temperature of the airplane's surface to above freezing. During the takeoff roll, one of the passengers saw droplets of water on the wings and yells; there is ice on the airplane. However, as our speed increased the water droplets quickly blew off leaving the wing clear of any ice. I heard him yell, oh it's coming off, and all went silent.

I filed a flight plan with the FAA so that we could refuel at Flower Aviation in Salina, Kansas. When you fill a jet with jet fuel at Flower Aviation, they give you six premium Kansas City steaks for free. Also, they have great snacks for the pilots and passengers.

In Salina, for some unknown reason, I felt I had to fly the airplane back to Phoenix. So I told Hal I was going to fly the airplane back. He had no problem with this, so I climbed into the left (pilot) seat and Hal in the right seat (co-pilot). We departed Salina and flew uneventfully to Sky Harbor Airport.

On the approach to runway 26 Right (west landing), I lowered the landing gear lever to the down position, and the sound from the hydraulic system delayed for about three seconds. Delays of this nature are not unusual, and even if the landing gear did not work, I have lowered the landing gear of this airplane numerous times under a training environment teaching simulated emergency conditions, I can do it without thinking. So, this delay meant nothing to either Hal or me.

I approached at the normal speed of one hundred and twenty knots (138 miles per hour) and again nothing unusual happening. I touched down smoothly, lowered the nose wheel to the runway and applied thrust reversers. On the rear of both engines are the thrust reversers. Metal doors that hydraulically deploy divert the engines thrust forward, opposite of the aircraft's movement forcing the airplane to slow down without using brakes. When the aircraft slowed to sixty knots (70 miles per hour) Hal announced

"60 knots," and I retracted the thrust reversers which must be brought back to the retracted position to keep from ingesting foreign debris from the runway into the engines causing damage.

Since our hanger was at the far end of an eleven thousand four hundred and eighty-nine feet runway, I let the airplane coast staying on the runway. I didn't apply brakes until I felt the airplane unexpectedly start to accelerate which was very unusual, so I applied brakes, but my pedals did not move. Under normal conditions, brake pedals move approximately one-quarter travel before feeling resistance, and you can feel the airplane start to slow down.

The normal hydraulic system on a Sabreliner produces three thousand pounds of pressure. The main brake system has anti-skid which prevents the wheels from locking when brakes are applied. Small generators installed in the wheel's axels sense tire speed. When the generators detect the tire speed going too slow, they release hydraulic pressure preventing the tires from skidding. A tire skidding and bursting during landing would cause the aircraft to lose directional control.

I have investigated enough aircraft accidents to know when someone loses control of an aircraft, the percentage of lives lost and severe injuries increase dramatically.

When I attempted to apply normal brakes, the pedals would not move. They froze in the full up position. I told Hal, "I don't have any brakes, you take the airplane." Hal applied his brakes, but his pedals were also frozen. I then took back control of the airplane and reached for the red emergency brake handle, but since we were unable to move the brake pedals, emergency braking was also unavailable.

When you pull the emergency brake handle, you redirect hydraulic fluid from its main source to a small container holding less than a gallon of fluid. Then, you must start pumping the brake pedals to build up brake pressure and hopefully, reduced braking will appear. You have to be very careful not to apply too much brake pressure because you don't have anti-skid and this could cause a tire to rupture.

Appling so much pressure on the brake pedals; I buckled the cartilage in my right knee and had to have orthoscopic surgery. Those pedals didn't move one iota.

I thought about pulling the landing gear up so that the airplane would collapse onto the runway, but again we would be out of control. I thought about ground looping ²the airplane, but you would have to apply full rudder deflection along with associated brake. Since we had no brakes, this maneuver was out of the question. Rudder alone would not be enough to ground loop the airplane. I attempted to redeploy engine thrust reversers, but the thrust reverser levers were frozen. I tried to shut the engines down but the throttles were also frozen, and they couldn't be moved rearward into the cutoff position. I attempted a go-around by applying full throttle, but the throttles were also frozen.

I looked beyond the runway's end and saw what looked like pipes standing upright. They were part of the Instrument Landing System Antennas. I didn't want the wings hitting these pipes for fear of splitting the fuel tanks open and atomizing the fuel. Atomized fuel is a spray that if ingested into an engine would cause it to explode. So, I turned off the right side of the runway and in between parked aircraft. Throughout all this maneuvering, the aircraft kept accelerating. This added speed, allowed me to raise the nose so that the bottom of the airplane would be exposed to take the impact of anything we hit. As we approached the perimeter fence on Twenty-fourth Street with the nose in the air, the fence went down without us feeling any impact or slowing.

With the nose up I could see across the Twenty-fourth Street and saw two light standards in the public parking lot perpendicular to the path we were on. I turned the aircraft more to the right and positioned the path of the aircraft to go directly between the two light standards. We went through the fence across the twenty-fourth Street; hit a six-foot concrete block wall knocking it down. Both wings then hit the two light standards exactly where I wanted and that brought the airplane to a stop.

Both Hal and I did not have even the slightest injury. I started shutting systems off that I felt could cause a fire. Since the engines were still running but with the throttles frozen, I had to close the fuel valves electrically to shut the engines down.

We heard the passengers yelling they couldn't get the main cabin door open. Hal calmly got out of his seat pushed the passengers back out of the

---

²Ground looping is when the aircraft starts spinning around on the ground and will cause the landing gear to collapse.

way, closed the door again, removed debris from under the door and opened it. They all exited the aircraft. I shut everything down, and as I was climbing out of my seat, I saw burning fuel running by the left side of the airplane. I yelled to the rear passengers to get out the emergency exit over the right wing. I hesitated to make sure they got the door opened, and then I exited through the main cabin door running through burning fuel.

I got to the front of the aircraft and stopped everyone, counted heads to make sure everyone was there and then turned and ran away from the airplane. Two individuals came running up to the airplane wanting to help, and when they saw everyone exiting the aircraft, they became so dumbfounded they couldn't move. We had to drag them away from the burning wreckage. When we got about one hundred feet from the airplane, it exploded into a ball of fire.

I have thousands of hours teaching all models of the Sabreliner. I have had my share of emergencies and have landed utilizing emergency brakes more than once. We taught students to use the emergency braking system to let them feel and understand its characteristics. All Sabreliner operators know that if you cannot move the brake pedals you not only cannot use normal braking but emergency braking is also inoperable.

A few months before this accident, an FAA Sabreliner was practicing emergency braking and blew two tires by applying brakes too hard locking up the wheels and going out of control. Fortunately, no damage had occurred.

Hal, being a doctor, checked everyone and not one person was injured in any way other than my knee. However, I lost my brown leather bomber jacket I had made for me special when I was in Korea. It was nothing less than heartbreaking, but I wasn't going back to look for it.

All controls for the throttles, brakes cables and rods, and thrust reversers go through the center pedestal between the pilot seats. If anything hard, like a wooden board, gets in there, it would become lodged and freeze all these controls. The fire would naturally destroy any evidence. Both Hal and I believe this was again an act of sabotage. However, the FAA didn't want even to consider sabotage because they wanted to pursue certificate action against me.

A photographer and reporter from the TV Series "COPs" were on the scene filming. They didn't interview anyone that I could see.

We hung around the wreckage until we got a ride from a city van to the Cities Corporate Terminal. I called the National Transportation Safety

Board's (NTSB) in Los Angeles and told Jim Erickson about the accident. I also told the FAA's Aviation Safety Inspector, Jim Kerr about the accident and then we all went home.

The following day all of us went back to the wreckage to see if we could help with the investigation. The NTSB and FAA were busy tearing into the wreckage trying to determine what went wrong. One of the passengers and his two sons talked to the local news channels, but I didn't want to give any statements, so Hal and I left.

Hal and I, at our own expense, traveled to Washington, D.C. at the request of the NTSB to help in their investigation. They wanted us to identify the various sounds recorded on the cockpit voice recorder and explain Hal's comment about the lowering of the landing gear hesitating before it actuated. Listening to the recording, we identified Hal when he was reading the landing checklist and the landing gear activating during the impact. Hal and I in perfect unison, yelled, "Holy shit" as the airplane hit the six-foot concrete block wall. We laughed about it afterward.

During this meeting with the NTSB, Hal and I made another statement as to the condition of the brake pedals, expecting them to include our statements in our report. Somehow, they disappeared.

The final NTSB accident report became public approximately nine months after the accident blaming both Hal and me for everything. The language they used was not specific at all but used general terms which could mean anything. My comment to their findings was, "The NTSB had nine months to find out what I could have done differently to save the aircraft," and I had only about twenty-five seconds. They also made the statement that the anti-skid system had failed. Another lie stated by the FAA because the aircraft's anti-skid system had been tested fully functional by its manufacturer.

The Division Manager of Flight Standards at the Regional Office in Los Angeles, Tim Forte transferred from the FAA to the NTSB in Washington Headquarters. He was one of the individuals who made the final approval of the findings of every aircraft accident reported to the NTSB. The NTSB accident database dates back to April 12, 1962. Our accident occurred on November 6th, 1991. I attempted to locate the accident report, which according to the NTSB accident database should have been there but it was missing.

It became obvious the FAA was again trying to satisfy their vendetta against me, and when I attempted to appeal their decision, they refused to even listen to the facts of the case, and their findings remained unchanged.

The FAA's vendetta against me began In May of 1983 when an Air Traffic Controller made a grave error in handling traffic at the Van Nuys Airport, California killing my brother George and his son-in-law Vince. My family sued the FAA and collected one million dollars for Vince's death. T the time, I was an aviation specialist at the FAA Regional Office, in Los Angeles, California at the time and I had taught Vince how to fly and trained him for his instrument rating. Even though I stayed out of the accident investigation and didn't participate in any way, the FAA blamed me for the lawsuit and began a vendetta against me that has lasted until the date of the book.

The FAA's local District Office investigating the accident interviewed me and agreed with my reasoning that emergency braking wasn't available because we could not depress the brake pedals. When you use emergency braking, planning becomes a key factor. It takes additional runway to stop, and when pumping the pedals, care must be exercised to prevent locking a wheel and bursting a tire causing the loss of aircraft control.

I knew they wouldn't work so why even take the time to pull the handle. The investigating FAA Inspector, Jim Kerr, passed this information up to the Regional Office in Los Angeles. They disregarded his report and initiated enforcement action against me from the regional office. They wanted to take my pilot certificate for forty-five days. Naturally, I appealed their decision, and eventually, it went before the National Transportation Safety Board hearing judge. When I appealed the FAAs proposed certificate action against me, a friend who retired from the FAA as a regional attorney that I must send a copy of my appeal to the attorney at the Western Pacific Regional Headquarters, Mr. DeWitt Lawson, which I did. My mistake was not sending it "Return Receipt Requested" because it disappeared also.

When the appeal started, it is done informally between the FAA and me communicating procedurally by mail with the NTSB Judge. The FAA immediately tried to get the NTSB judge to award them the case because, according to the FAA, I never notified them of my attention to appeal. In other words, my letter I sent notifying them of my appeal mysteriously disappeared because I did not send it "Return Receipt Requested." The devious ways of the FAA are exposed.

Conspiracy

After a few weeks of examining the evidence, the NTSB Judge dismissed the case and stated I should have gotten an award for what I did. I told the FAA, "FAA nothing, Adams one (won)" but they didn't see the humor in my statement.

The TV Series, COPs just happened to be in the area riding in a Phoenix Police Car when the call came in, and they took a video of the airplane as it burned in the airport parking lot. Cops aired what footage they had on TV, and it has rerun many times. COPS contacted Hal and asked if he and I would be willing to reenact the entire flight for one of their episodes. Hal turned them down. Without Hal, I couldn't expect the producers of COPS to agree to use only me.

This accident occurred on November 6, 1991, and Hal and I have been the closest of friends ever since. We always played golf together, and at his request, I promised not to write about this accident or tell anyone about our thoughts on whether this was an act of sabotage. For the past twenty-four years, I've kept my word. Dr. Hal Price passed away in November of 2015, and I'm devastated by the loss of such a great man and friend. However, I now feel I can tell what happened because it has greatly affected my life.

The following page is a copy of the NTSB's final accident report. In their analysis, the NTSB stated, "The co-pilot lacked experience in the aircraft and crew coordination during the approach landing, and the emergency was ineffective."

Federal Aviation Regulations demands an individual to attend formal training to be qualified as a co-pilot in a jet airplane. Dr. Price satisfactorily completed this training and had flown numerous hours both as a pilot and co-pilot in his Sabreliner.

So, this statement from the NTSB that Hal lacked experience is a pointless statement and contains nothing worthwhile. Also, I am an FAA certified flight instructor in Sabreliners, and I gave Hal more instruction than that required by the Federal Aviation Regulations.

This statement is hollow and means nothing. Did the FAA have a ghost in the cockpit along with Hal and I, to determine there was a lack of crew coordination? What did they consider the co-pilot's lack of experience? The cockpit voice recorder contained the final check we completed before landing. Their statement was an attempt to discredit us to make it appear as if we didn't know what we were doing, thereby supporting the FAA's violation against me.

The NTSB report ignored a key factor. When I reported the accident to the NTSB the night it happened, they were told by me that the brake pedals were frozen in the full up position thereby rendering both normal and emergency brakes inoperable. Their report did not contain either mine or Hal's statement referencing this key factor. They just ignored these statements and didn't even reference them. Admitting this would have removed the FAA's justification for attempting to suspend my pilot certificate. This factual evidence proves the NTSB falsified their report to appease the FAA.

Also, the NTSB's accident report determines the probable cause(s) of this accident to be: THE DELAY OF THE PIC (Pilot In Command) TO APPLY NORMAL BRAKING AND HIS FAILURE TO EXECUTE THE APPROPRIATE EMERGENCY PROCEDURES CONTRIBUTING TO THIS ACCIDENT WAS AN UNDETERMINED ANTISKID MALFUNCTION.

Again, this is another FAA/NTSB lie. The anti-skid manufacturer tested the system on the aircraft and found it operated normally. Also, with four thousand feet of runway remaining and the speed of the aircraft at sixty knots, delaying the application of normal braking would not affect the aircraft's ability to stop well before the end of the runway had the brakes operated normally. Once again, the NTSB provided the FAA with a fabricated accident report which added to their vendetta against me by removing the two witness statements identifying the cause of the brake failure.

When first hired by the FAA, I attended two classes conducted by the NTSB Accident Reporting School in Oklahoma City, Oklahoma. They were very adamant about how to determine the probable cause of an accident. When listing probable causes, if you remove a probable cause and the accident wouldn't have happened, then that probable cause is valid. However, if you remove the particular causal factor and the accident still occurred, then it wouldn't be considered a viable casual factor.

The NTSB is supposed to be an independent organization designed to render decisions independent from the FAA Like an overseer. However, when Hal and I went to Washington, D.C. to assist in their investigation, I met the then Division Manager of the FAA Western-Pacific Region, Tim Forte who is now deceased. He had just transferred to the NTSB and who has the final decision in determining Probable Cause of this accident. It then became clear who fabricated the Cause and Findings of this accident in an attempt to place the blame on me.

Conspiracy

I contacted the NTSB Investigator who was in charge of investigating the accident and wrote the final report. I asked him why he removed the statements made by Dr. Price and myself explaining that the pedals were unusable because of being frozen in the full up position. His stated he submitted our statements with the final accident report to Washington Headquarters but were removed after it got to Washington Headquarters.

National Transportation Safety Board
Aviation Accident Final Report

Location:	PHOENIX, AZ	Accident Number:	LAX93FA033
Date & Time:	11/07/1992, 2226 MST	Registration:	N169RF
Aircraft:	Sabreliner Corp. N-265-60	Aircraft Damage:	Destroyed
Defining Event:		Injuries:	8 None
Flight Conducted Under:	Part 91: General Aviation - Personal		

Analysis

UPON LANDING AT THE COMPLETION OF A CROSS COUNTRY FLIGHT, THE CAPTAIN OF THE TURBOJET AIRCRAFT EMPLOYED AERODYNAMIC BRAKING AND THRUST REVERSE TO SLOW THE AIRPLANE TO ABOUT 60 KNOTS. THE CAPTAIN WAS ALLOWING THE AIRPLANE TO ROLL TOWARD THE END OF THE RUNWAY WHERE THE OWNER/CO-PILOT'S HANGER WAS LOCATED. WITH ABOUT 4,000 FEET OF RUNWAY REMAINING, THE CAPTAIN APPLIED THE BRAKES. NO BRAKING ACTION WAS NOTED. THE AIRPLANE CONTINUED OFF THE END OF THE RUNWAY, THROUGH A FENCE AND BLOCK WALL INTO A PARKING LOT WHERE THE LEFT WING OF THE AIRPLANE WAS SEVERED. A POST CRASH FIRE CONSUMED ABOUT HALF OF THE AIRPLANE. EMERGENCY BRAKING PROCEDURES WERE NOT EMPLOYED. THE CREW REPORTED THAT THE WERE UNABLE TO SHUT DOWN THE ENGINES. THE CO-PILOT LACKED EXPERIENCE IN THE AIRCRAFT AND CREW COORDINATION DURING THE APPROACH, LANDING, AND EMERGENCY WAS INEFFECTIVE. THE AIRPLANE TRAVELED ABOUT 11,000 FEET FROM POINT OF TOUCHDOWN TO POINT OF REST. EXAMINATION OF THE BRAKING AND HYDRAULIC SYSTEMS FAILED TO PINPOINT A MALFUNCTION.

Probable Cause and Findings

The National Transportation Safety Board determines the probable cause(s) of this accident to be: THE DELAY OF THE PIC TO APPLY NORMAL BRAKING AND HIS FAILURE TO EXECUTE THE APPROPRIATE EMERGENCY PROCEDURES. CONTRIBUTING TO THIS ACCIDENT WAS AN UNDETERMINED ANTISKID MALFUNCTION; THE COPILOT'S INEXPERIENCE IN THE AIRCRAFT; AND INADEQUATE CREW COORDINATION.

FIGURE 23 - COPY OF NTSB ACCIDENT REPORT PAGE ONE - CAUSE AND FINDINGS

Findings

Occurrence #1: AIRFRAME/COMPONENT/SYSTEM FAILURE/MALFUNCTION
Phase of Operation: LANDING - ROLL

Findings
1. (F) LANDING GEAR,ANTI-SKID BRAKE SYSTEM - FAILURE,TOTAL

Occurrence #2: ON GROUND/WATER COLLISION WITH OBJECT
Phase of Operation: LANDING - ROLL

Findings
2. OBJECT - FENCE
3. (C) BRAKES(NORMAL) - DELAYED - PILOT IN COMMAND
4. (C) EMERGENCY PROCEDURE - NOT PERFORMED - PILOT IN COMMAND
5. (F) CREW/GROUP COORDINATION - INADEQUATE - PILOT IN COMMAND
6. (F) LACK OF FAMILIARITY WITH AIRCRAFT - COPILOT/SECOND PILOT

LAX93FA033

FIGURE 24 - COPY OF NTSB ACCIDENT REPORT
PAGE TWO - CAUSE AND FINDINGS

Chapter 8

ARRIVA AIR INTERNATIONAL

I met Steve when he called me to ask if I could help him start up a new cargo-only air carrier company under Federal Aviation Regulations 121.

He was a medically retired Phoenix Police Officer. He received his injuries while chasing a subject on foot when he slipped and fell dislocating his leg at the knee causing it to be ninety degrees from its normal position. He received a medical disability from the department. We met at the Phoenician Restaurant to discuss the possibility of linking together to set up the new an air cargo operations out of Williams Airport in Mesa, Arizona.

He would be president, and I would be Vice President/Director of Operations. It sounded good, and we made a verbal agreement.

In the meantime, I was a freelance test pilot for Dace Kirk, who owned a company called Phoenix Composites that built custom airplanes. My job was to test fly the finished product and then train the owner. I was paid $125 per hour for both testing and training. Phoenix Composites specialized in building a composite (fiberglass) Glasair III airplane, which is a high performance, single engine low wing aircraft. It weighed just 1700 pounds and had a high-performance Lycoming engine developing three hundred and sixty horsepower. When you build an experimental aircraft and its ready to fly, it must remain within a small area of airspace until it accumulates forty hours of flight. For engine break-in the first ten hours are spent directly over an airport, flying racetrack patterns in case the engine failed, it could be landed it without causing undue hazards to the airplane, pilot or anything on the ground. Engine failures were not unusual for numerous reasons, the number one being fiberglass clogging the fuel pump.

I two individuals from Big Bear, California contacted me saying they had just bought a Glasair III from the Musician Kenny "G," and could I train them in their airplane? I agreed, and they showed up in Mesa, Arizona in their newly purchased airplane. I trained them for a week. One of the idiosyncrasies in this particular airplane is when you stall [3]this airplane it will immediately become inverted so fast it will take you by surprise.

To recover from a stall, you lower the nose until flying speed is again reached. When this aircraft stalled, it immediately rolled upside down, and the student pushed the control to lower the nose which is correct. However, we turned upside down so quickly; it felt as if the student was executing an outside loop. He immediately corrected the problem, and everything turned out OK.

I test flew a Glasair III, built by Lynn Babcock and Manny Ramirez. Lynn was a property developer and rented warehouse space to manufacturers in the Scottsdale Airport complex. After the airplane testing was over, I performed all the aerobatic maneuvers the airplane was approved for and signed the log book showing each maneuver I performed.

While training Lynn in his Glasair, we were flying near Ryan Field in Tucson, Arizona when the engine quit. The airplane was out of the test area and had accumulated over forty hours. This engine failure surprised me because normally engine failures occur within the first ten hours of flight because fiberglass residue accumulates in the fuel pump starving the engine of fuel. However, this was not the case. The engine was surging, and then it just quit. I immediately told Lynn I had the airplane. I took control and turned toward Ryan Field (airport). The Glasair III was a high-performance aircraft but didn't have a very good glide ratio. When I talk about the airplanes glide ration I use the expression, "It glides like a grand piano."

We were coming down fast, and Lynn radioed the control tower and told them we had an engine failure and were attempting to land on Runway 6 (northeasterly direction). The controller ordered all other aircraft away from

---

[3]Is when the wing of the aircraft quits flying because of a disruption of wind flow across the wing.

runway 6 and cleared us to land. The aircraft was descending very rapidly, and it appeared as if we were going to land short of the runway. However, I kept the landing flaps up which reduced our rate of descent slightly. [4]When we got close to the ground, I gradually raised the nose of the aircraft until our rate of descent stopped, and we were able to extend the glide over the Saguaro Cactus' until reaching the runway. We landed safely. After stopping we had to push the aircraft off the runway until a vehicle could come and tow us to the parking area. Lynn and I have been friends ever since.

Meanwhile back at Arriva, Steve made contact with a Japanese firm in Tokyo that had just bought a Boeing 727-200 cargo configured airplane located in Fort Lauderdale, Florida. It had just been rebuilt from stem to stern and converted from a passenger to a cargo configuration with new flooring containing small wheels for positioning cargo containers. The Japanese company that owned the airplane wanted to lease it to us.

Arriva Air, Inc. had already bought an FAA Operating Certificate allowing us to operate this aircraft internationally. Arriva Air was in the process of obtaining a contract from a Taiwan firm hauling Tuna Fish (Sashimi) from the Island of Palau in Indonesia to Tokyo, Japan. Steve and I went to Palau to look the operation over and spent a week on the island.

We flew Continental Airlines to Hawaii, Guam, then to Palau. Approaching Palau, the pilot flew down low over the area to show everyone the remnants of Japan's deteriorating equipment from WWII. The runway is coral rock and wears out more quickly than an asphalt runway.

When the captain flew around the area, numerous small islands were in the area having vegetation growing making the island look like a mushroom. During WWII, the Japanese used to hide their seaplanes under the overhang to protect them from being seen from the air. Heavy jungles covered the islands. Throughout the jungles, we saw rusted tanks, trucks, and corroded airplane parts the Japanese left when the war was over. It was interesting to see how much war material was on the island.

---

[4]Landing flaps are at the trailing edge of the wings and when lowered allows the speed of the aircraft to be safely reduced for landing.

The Island of Palau was a vacation resort. We ate dinner at the local hotel, and the company we were dealing with had two representatives who enjoyed their scotch whiskey.

We visited the piers where the fish would be taken off the boats, tested by inserting a tube into the fish and extracting a meat sample. The fish was then packed in ice and transported to the airplane. The airplane would then fly to Japan supplying fresh sashimi (raw fish) to the markets. The fish had to be delivered to Japan the same day it was caught to keep it fresh.

The only communication on and off the island was by two-way shortwave radio. No telephone service was available because of the island's remoteness. Steve and I went to the harbor to see what facilities were there. We walked to the edge of the dock and looked down into the water. To our amazement, we could see about sixty to seventy feet to the bottom. The water was so clear it made you feel it was ok to drink, but we knew it wasn't because it being salt water.

We reluctantly left Palau and came home through Guam and Honolulu again. It was a beautiful trip, but it was good to get home again.

Dace Kirk was waiting for me to test fly a Glasair III he had just finished, so I spent the next week going around in circles above the airport. Once it was out of the test area, I flew it to Dallas, Texas to the owner. I stayed there to train the new owner, and he was ecstatic with his new airplane. He did a good job flying, [5]and I left to come home.

Steve, Skip the Chief Pilot, and I had a meeting because there were problems with the Air Carrier Certificate Steve had bought from a company in San Jose, California. Everyone at Arriva, along with the FAA decided the best course of action would be to surrender the certificate to the FAA for cancellation. In doing this, the FAA made a promise they would give us priority in obtaining a new one. However, the FAAs vendetta against me came alive once again, and they delayed keeping their promise, and we waited anxiously without the FAA doing anything.

Skip, and I flew to Fort Lauderdale to pick up our Boeing 727 and bring it home. I hadn't flown a B- 727 since I was with Eastern Airlines but

---

[5]A Boeing 727 has three engines on the tail end of the aircraft and was very popular carrying airline passengers.

Conspiracy

we had a ball flying it back to Williams Airport. Upon arriving, I did a flyby at very low altitude but Skip made the landing. Everyone was there to greet us. It was like we were arriving for a party.

Steve took all the company's operating manuals and boxed them up for shipment. Steve asked me to deliver them personally to our potential customer who was going to set up a fishing operation using our B-727. Steve had made arrangements for me to meet this customer in Manila, Philippines.

When I arrived in Manila I the person I was supposed to meet never showed up. I then received word that he never left Taiwan and asked if I could fly to Taiwan and meet him there. So, after two days of waiting in Manila, I boarded Philippine Airlines and flew to Chiang Kai-shek Airport in Taipei, Taiwan.

When I got to the hotel, I came down with the flu, and for three days I laid in bed sick. No one came to get the manuals I had carried all the way from Phoenix to Manila to Taiwan. When I felt better, I got in a taxi to go to the airport to return home when a car started following us. The driver of the car I was in saw it immediately. Finally, that car passed us on the left and forced our car to the curb. The driver got out, and I was ready to fight, but all he wanted was to take charge of the manuals and to take me to my airplane. Whew!

When I got back to Williams Airport, I found that copies of the Air Carrier Certificate we had surrendered to the FAA was hung on the wall of the office as if they were still in effect. When customers showed up, they saw the certificates on the wall and did not realize they were just copies of the originals that the FAA received for cancellation.

They inspected the B-727 not realizing it wasn't legal to fly because of outstanding Airworthiness Directives on two of the engines. They believed these items proved the company was legitimate and issued a Cashier's Check for three hundred and fifty thousand dollars. I do not know whatever happened to that money because I was leaving the U.S. heading for Saudi Arabia.

When I found out what had happened, I immediately resigned and made application to the United Nations as an Aviation Safety Inspector for the International Civil Aviation Organization (ICAO). My wife Nancy wanted a divorce because she couldn't take the pressure the FAA was putting on me any longer. The FAA had been trying to suspend my pilot certificate for forty-five days. Even though I had beaten the FAAs plan and put all this behind us, she couldn't take it any longer, so we split up, and I couldn't blame her. Chalk up another one for the FAA.

Chapter 9

SAUDI ARABIA

I received confirmation from the International Civil Aviation Organization (ICAO) of the United Nations that I had was selected for an aviation position in Jeddah, Saudi Arabia. Within two weeks I was on my way to ICAO headquarters in Montreal Canada to meet everyone and to get special instructions before going to Jeddah. When I got to Montreal, it was snowing heavily. It was the first time I had been in snow in a long time.

I spent two days in Montreal then boarded an Air Canada flight to London, England. Eight hours later I arrived in London where I was to spend the night before heading to Jeddah. It was my first experience in an English Pub and their warm beer. I loved it. Normally warm beer in the U.S. doesn't taste good. The British beer is delicious. They put a head on each beer that doesn't dissipate when standing. It was creamy and very tasty.

I arrived in Jeddah late at night, and Don, the Director of the local ICAO office, met me as I was leaving customs. I later found out that Don was one of the prisoners held hostage by the Iranians for fourteen months.

They took me to the compound where I would be living and allowed me to go right to bed. A compound is nothing more than many apartments with a wall around it. In Saudi Arabia, there is nothing strange about an apartment complex or a single house having a ten-foot wall around it. It was the first week of Ramadan (Muslim) which lasts a month. For the first week of the holiday our office closed. Throughout the entire month of Ramadan, everyone fasts until the evening. Then they gorge themselves.

Conspiracy

Weekends in Saudi Arabia are Thursday and Friday, not Saturday and Sunday. Since the office closed for the first week of the holiday, it would work out fine because it would give me more time to get settled in my apartment.

The next day I met the individual who will be training me. He was still an FAA Aviation Safety Inspector from Kansas City, Missouri. He had gotten in trouble for something he did in Jeddah, and was being relieved of duty and sent back to the states. I never asked what kind of trouble he got into because it was none of my business.

Three weeks later he was on his way home. His replacement was an American from Florida and who was an ex-Saudia (Saudi Arabian Airlines Special Flight) pilot from across the airport. As new as I was, I trained this new guy Jim Long because he had no previous experience as an Aviation Safety Inspector. It didn't matter to me; I was just happy being away from the FAAs vendetta.

Jim and I traveled around Jeddah, him showing me places he felt I should know about because he had been living in Jeddah for seventeen years. One day we ran across a motorcycle shop selling Harley Davidson Motorcycles. We stopped, and I bought a new Harley Roadking and joined the local Harley Owners Group (HOG). The HOG was a group of motorcyclists who got together on weekends and road all over the place. The members of this group were from all throughout Europe, and only a few were Saudi Arabians. They were a great bunch of people out to enjoy themselves just riding. Harley's motto is "Born To Ride-Ride to Live," and that's what we did.

My first outing with them was in the local mountains where the owners of the Harley Shop set up an open tent and served breakfast. It was great fun and gave me a chance to meet the members. One of the riders I met was Yassah Bin Laden, brother of Osama Bin Laden but much shorter. Yassah was only about six foot where Osama was supposed to be six five.

I was told never to mention Osama's name out of respect for Yassah. Osama was his brother or half-brother because his father had four wives. In the Muslim community, the man is allowed, four wives. Each wife must have her own home equal to that of the other wives. It was easy to identify a man with four wives because you would see four identical houses alongside one another.

Yassah had a great sense of humor. One day I walked into the Harley Shop, and Yassah was sitting at the coffee bar. He called me over and asked if I was going to the American Counsel grounds to give a ride to anyone wanting to ride on a Harley. I asked what the occasion was and he said they were having a Halloween Party. I asked what kind of costume he was going wear and he said he intended dressing up as Osama. I burst out laughing because we had been trying not to mention Osama's name and here Yassah says this. Yassah was a lot of fun, and I enjoyed being around him. Yassah owned Bin Laden Construction Company worth five billion dollars.

I met the Chief Pilot for the Saudi Arabian's National Wildlife Refuge located in the mountains near Taif City. He was an American living at the refuge and had been there ten years. He was telling me about their attempts to bring back the Antelope population. The Saudi's killed many of the Antelope just for the sake of killing. They almost caused the antelope to become extinct in their country. This organization has been very successful, and he invited the HOGs to come and visit to see their progress. I posted this invitation on the bulletin board at the Harley Shop. Fourteen members signed up.

FIGURE 25 - MOUNTAIN ROAD TO TAIF

We left the Harley shop headed for Taif, but before we even got out of Jeddah, the police stopped us. I think they just wanted to look at all the bikes. Motorcycles in Jeddah were a rarity and seeing fourteen at one time was unusual. It was also illegal because anything over five is considered an illegal gathering.

There were four members of the HOG that were Saudi, so there wasn't a language barrier. After about thirty minutes we were allowed to be on our way. We were again stopped and directed to continue straight ahead because this was the exit to Mecca and no infidel was allowed in that city. Most of us in the group were infidels (non-believers).

We continued to Taif, and when we got to the city line, the police had set up a roadblock. For the third time, the police stopped us for no apparent reason. I laughed and told everyone it was payback for me because when I was a cop in Los Angeles, we stopped the Hells Angeles every time they came into our district they were escorted out.

We stayed with the Police for about an hour while they figured out what they were going to do with us. In the meantime, we decided to eat in this small restaurant with dirt floors. The Saudi bread with hummus is delicious, especially when it's freshly baked. This restaurant had its ovens, so the bread was very fresh.

After eating we went back to where the police were watching our bikes and asked if they made up their minds on what to do. They said "yes," and with red lights and sirens, they escorted us all the way to the wildlife refuge.

We went on the tour and saw the antelope herds and a rare Saudi Arabian leopard. They were looking for a female to mate with the male they had in captivity. A beautiful animal, but its color did not appear to be as bright an orange as an African Leopard. The color was like the Leopard had a dark shadow over its body, but it was still beautiful.

We left the refuge to return to Jeddah. When we went out the main gate, the police were waiting. Red lights and sirens blaring, they escorted us through town and out the other side of Taif and down the mountain. I wanted to stop and take pictures of the wild baboons that were running around, but the police wouldn't stop. Every so often these baboons would go through the city creating havoc becoming a big nuisance. Signs posted, warned people to keep their arms inside their vehicles and don't try to feed them. People have lost arms because the Baboons are very strong and if anyone tries to feed them they could lose their arm.

About two weeks after returning from Taif, Yassah Bin Ladden came up to me in the Harley Shop and gave me a badge with the inscription "Al Adams Jeddah Saudi Arabia" embedded in the metal in blue plastic lettering. Each member of our group to Taif got a badge. I now have it on my leather vest, and when people see it, questions arise.

FIGURE 26 – BADGE GIVEN TO ME BY YASSAH BIN LADEN

There was a Sudanese Secretary assigned to work with the personnel at General Aviation Kingdom of Arabia (GAKA). He wasn't assigned any particular unit but worked wherever needed.

He and I were talking, and he asked if I had been to the grave site of Eve (Adam and Eve) in the downtown area of Jeddah. He explained the exact location and I remembered seeing the only area filled with plants and trees. It was one block from a white marble Mosque we called "Chopping Block Square."

I told him I knew where he was referring and I said I was going there to take some pictures. [6]He got upset and told me I couldn't do that, and I replied, "Why not, that was my mother too." He laughed and said he never thought about it like that. All forgiven.

Chopping Block Square is a white marble Mosque located on a beach at the end of an inlet from the Red Sea. Every Friday, which was open to the public, condemned prisoners, would be executed by the Saudi Arabian government by cutting their heads cut off using a very broadsword.

---

[6]Muslims do not want their picture taken because they feel their soles will be in the camera.

Conspiracy

As a teenager, I remembered seeing a picture in Life Magazine, of a man beheaded when Fidel Castro took control of Cuba and I never wanted to see anything like that again, and I haven't.

We had a new airplane mechanic arrive along with his wife. They were living in the same compound as the pilots. The Gulfstream IV representative from the Savannah, Georgia took both of them to Chopping Block Square to watch the prisoners beheaded and when they returned they had nightmares for two weeks.

A Gambia Airlines pilot came to my office and understood I wrote technical manuals. I said .yes, and he asked if I could write a training manual for pilots for a Boeing 727 and a 707. I told him I could handle that without a problem. He then asked if I could write a training manual for flight attendants for the Boeing 707 also. I told him yes but that it would cost five thousand dollars for each manual. He told me, "No problem, I'll be back tomorrow." I didn't think I would ever see him again.

The following day he showed up at my office with a shoe box under his arm. He gave me the shoe box and in it were stacks of one hundred dollar bills. I almost choked. The next day I went and bought another Harley. I almost sold it because it was a new model that had just come out and it was one of two in Saudi Arabia. I was offered two thousand dollars to sell it but declined the offer.

Every night I worked on those manuals until early morning hours. I had them done within two weeks. I called and gave him the manuals. He immediately took them to the air carrier section of the Saudi GACA for approval. Within a week all three manuals were approved. I never told anyone I wrote those manuals. I left things alone so as not to stir up a hornet's nest.

Saudi Aramco called and reported they had a fatal accident in the Persian Gulf and that fourteen people lost their lives. My boss, Mohammad Kahn was a retired Brigadier General from the Saudi Military. He specialized in helicopters when he was in the military. So when this report of a helicopter accident came in, he immediately wanted to be involved. However, he also informed me I was going with him to Saudi Aramco in Dammam located on the Persian Gulf. We arrived at the airport, and the Aramco chief pilot met and drove us to their hangar. We looked at all the records for the Bell 414 twin turbine helicopter. I was asked to go to the Oil Rig where the helicopter departed and crashed. General Kahn stayed in Dammam. Fourteen passengers were on board the helicopter when it departed an oil rig and crashed into the Persian Gulf. Twelve passengers perished along with two crewmembers. Five passengers survived by kicking out and escaping through the entrance door window.

I talked to the witnesses, and they described the helicopter's takeoff. It lifted off the platform moved forward as if taking off and all of a sudden it nosed down rolled right and crashed into the water. To me, it seemed strange that the crew also perished because each pilot's station had oxygen that was operational underwater. The oxygen system held enough air to allow both pilots to evacuate the helicopter while under water. I wondered why they stayed in the craft and perished because both were still in their respective seats with seat belts fastened.

We had an interpreter listen to the cockpit voice recorder because both pilots were Saudis. The interpreter said the pilots did nothing but prayed throughout this tragedy and all he heard both crew members announce as they were dying, "God is good, and there is only one God." They did not try to correct the helicopter from nosing over and rolling to the right. Nor did they attempt to get out of the helicopter after it hit the water. The people in the back died because they couldn't break the Plexiglas window to escape the aircraft.

We analyzed the Flight Data Recorder and learned everything in the helicopter was working properly. The flight recorder had all the engine parameters. They showed both engines RPM had reduced substantially due to the pilot overpowering the engines because of taking off in a heavy helicopter into a tailwind condition. This tailwind forces the downwash from the rotor system back into the rotor blades diminishing its lifting capability.

The wind at the time of the accident was coming from the rear of the helicopter or in other words a tailwind. My summation along with General Kahn's was that the flight crew attempted to take off with a tailwind. If the wind was coming from the front of the helicopter, it takes less power to become airborne because the rotor system encounters fresh air whereas with a tailwind it takes a substantial amount of additional power to become airborne because it has to overpower its downwash. This reasoning is the tailwind blows the downward rotor wash forward and back into the rotor system causing the helicopter to lose much of its lifting capability. Couple this with the heavy weight of the passengers along with low rotor rpm created an unstable condition for the helicopter.

Conspiracy

The last thing Kahn said to me was, "You write the report." It was not unusual for I expected it. I didn't mind though because I liked Kahn a lot and it didn't bother me that I would be doing the report.

General Kahn was originally from Turkey, and he and I would talk a lot about my experiences in Istanbul. It was obvious he liked Turkey as I did. I made many friends there.

Net Jets Middle East had just gotten their air carrier certificate from the Saudi General Aviation Civil Authority (GACA). The acronym is pronounced GA…KA. The Saudi Arabia pilot license has a good looking emblem on it that was a very colorful piece of artwork. GACA issued me an interesting number. It made me a part of the James Bond legacy.

FIGURE 27 – PILOT CERTIFICATE ISSUED TO ME BY SAUDI ARABIA AUTHORITIES

The Saudi's hate Israel, and when they added all my aircraft qualifications to my pilot certificate, they left off the Jet Commander because it is now an Israeli aircraft. They transferred all my other qualifications without a problem except Jet Commander. I just laughed because it was no big deal. No Air Line flights originate in Saudi Arabia going directly to Israel. You have to go to another country first and then to Israel.

There was a Saudi air carrier inspector in our office named Ibrahim. He and I would laugh at anything and everything. What a great guy. Many

Saudi women wore black Abayas that covered every inch of their entire bodies. Ibrahim would tell me when he sees a women's ankle it would get him excited. I laughed. Some of the women would swim in the Red Sea wearing their Abayas, but they would not go in deep water that was over their heads because the weight of a wet Abaya would drag them down.

The Director of Operations for Net Jets Middle East came to me and asked if I would like to come to work for them as Director of Training. To entice me a little, they told me I would get qualified in the Gulfstream IV aircraft. This aircraft is considered the Rolls Royce of Corporate Aviation. Naturally, I said yes. A month later I moved out of the compound I was in and moved to a penthouse apartment in another compound where Net Jet Pilots lived. I had two patios on the roof, and the parties we had were fun. They also rented me a brand new Chevrolet four-door car for my transportation.

I went back to Savanah, Georgia where Flight Safety had their Gulfstream IV (G-IV) training facility. The Gulfstream IVs in Saudi Arabia had no chrome in the airplane; instead, it was eighteen karat gold. You would think it looked gaudy, but it didn't. It was beautiful. The only thing that did look rather gaudy was the gold toilet and sink. Those items were too big for that much shiny gold.

Not having flown an airplane for almost three years, my boss Frank told me not to worry if I flunked training. I could finish training when I returned for recurrent training after six months.

This course is three weeks in length, and I satisfactorily completed the course on time and with above-average grades, flying and in the classroom.

After about four months I returned to the US again and was sent to Kansas City Missouri to receive training and certification in the Beechcraft Hawker 800XP. I spent three weeks receiving training with another Net Jet Captain. He was a retired British Air Force Pilot, and this was his first experience flying the Hawker. I was hoping to take a Hawker back to Saudi Arabia, but it wasn't to be. General Ali Al Dhabi came to Kansas to enroll his son in the local university. Net Jet asked that I stay in town to show him around and to guide him through the process of signing his son up at Kansas City University.

Ali and I had worked together at GACA, and I liked him very much. His position with GACA was as Principal Operations Inspector replacing General Kahn when he retired. Ali and I got along fabulously. When we worked together at GACA, we would talk about the Muslim religion. It's origin with Mohammed and its history. He was very interesting. He retired as a Brigadier General from the Secret Police in Jeddah. Frequently, active military personnel would come into his office to greet him. I should say to kiss his rear end.

I spent a week with Ali until he went back to Saudi Arabia. I traveled to Savanah and then went back to Saudi Arabia. The Hawkers showed up while I was still in the US, but we took them out for a test flight. They were sure a wonderful airplane. That British Pilot told Frank, my boss and Director of Operations that I flew the airplane better than he. I think this upset Frank because I hadn't flown for almost three years before going to G-IV school. Frank didn't expect me to pass the flight training and he told me not to worry if I failed. I didn't fail but passed with compliments from my instructors.

FIGURE 28 - ENTRANCE TO THE VALLEY OF THE KINGS

FIGURE 29 - CITY OF LUXOR, EGYPT

In November I asked my oldest daughter, Stephanie, her soon to be husband Ryan and my ex- wife Linda to meet me in Cairo, Egypt. I must have had a screw missing, inviting my ex-wife but she enjoyed herself, and we got along fine. We all flew to Luxor on the Nile and boarded a shallow draft cruise ship and toured down the Nile visiting Aswan, Luxor, Valley of the Kings and the Cairo Museum. The last day on the ship a belly dancer came on board to entertain us. She asked for a volunteer to help dance with her. My daughter Stephanie went and assisted. Stephanie's ability to dance is unlimited, and the two of them were a sight to see. It was wonderful.

We visited Aswan Dam which prevents the Crocodiles from getting into the upper Nile. Before filling the reservoir with water, they had to move many three-thousand-year-old temples. They moved every building block by block to a new location above the expected water line and opened the area to tourists. However, the only way to get to this new location is by boat.

During one of my visits back home, I met a gal from Scottsdale named Suzi. We dated almost every night, and she wanted to come with me to Saudi Arabia, but they don't allow foreign women into their country that are not married. Suze kept bugging me about going to Saudi Arabia with me, but I kept telling her she can't. At her insistence, I finally agreed to marry her, and I told Greg Knapp, who was living in Vienna, Austria at the time, of our plans. Greg contacted me and told me we should get married in Salzburg, Austria in a real fifteenth-century castle. It sounded wonderful, so we made all the arrangements. I went back to Saudi and Suzi and her family along with another mutual friend, Greg Sharp who is a Pilot for America West Airlines made the arrangements to meet me in Vienna, Austria.

We all got together and rented a VW Bus, and we followed Greg and his wife Marsha to Salzburg. Greg had made reservations for all of us at the castle, and since it was out of season, we had the whole place to ourselves. It had just snowed about three inches the night before, and the entire area was beautiful, with the white snow on the ground, and the deep blue sky made quite a picture.

An American Minister who moved to Salzburg five years earlier married Suzi and I. The ceremony presided in a Chapel, built in 1520. There was a four-piece quartet; a harpsichord, two violins and a Cello that were playing before the ceremony. About halfway through, the minister called an intermission, and we all listened to the quartet play Mozart who was born in Salzburg. When their piece ended, the ceremony was resumed. We then had champagne at the reception in a restaurant built in the year 803. It is the oldest restaurant in Europe.

The following day we returned to Vienna with Greg leading the way. He was driving his BMW and the rest of us on the VW Bus. We entered this highway, and Greg increased his speed to over one hundred miles per hour. That VW Bus was becoming unstable, and I flashed my lights at Greg and then blew my horn. He finally slowed down his speed. His excuse was we were on the famous Auto-Ban, and he wanted us to experience the speeds that were allowed. It was a little unnerving for everyone in the bus that was not designed to travel at that speed.

The two days later, I left and returned to Saudi Arabia, and everyone else returned to Arizona. I contacted Suzi and told her I was going to start the paperwork for her to come over. Her reply was, "I'm not going to Saudi Arabia." Again, this was her answer when I called her the following week. I felt this is one hell of a way to start a marriage.

After I completed my contract with Net Jets, I left Saudi Arabia and returned to the US, bringing along my two Harleys. I wanted to concentrate on making this marriage work. I moved into my wife's house, and things just weren't right. As an example, I suggested I bring my Bose stereo system I bought in Saudi and hook it up in her living room. Her comment was, "you're not going to bring your toys into my house." After one month with her attitude like it was, I knew it wasn't going to last, so I walked out and got a divorce. She attempted to get a share of my house I owned in Vista, California but that was struck down because I purchased the house before we were married. We got the divorce.

By coincidence, I ran into a friend, Richard Castillo who introduced me to Mr. John DePalmer. John was a representative with IJet, LLC. John and I had known each other through Hal Price when he bought another Sabreliner. We talked for about five minutes, and he hired me on the spot as Director of Operations for IJet.

The operating manuals for IJet desperately needed revising. I started by rewriting some of the company manuals which I had to submit to the FAA for approval. I was nervous going into the district office because of the FAA's previous vendetta. Dick Thomas who became the office manager immediately asked me to come into his office. Once inside, he apologized profusely to me for what happened with the Gary Koch as manager of the FAA office and said he had no way of intervening over his boss, Gary. I knew Dick had nothing to do with the FAA's vendetta, but I accepted his apology anyway. It appeared the attacks on me were over, or at least I hoped they were.

As Director of Operations for IJet, I flew a Learjet 35 and a Hawker 800 on charters. The Learjet operation had a pilot, Rick Sprout, who retired from Continental Airlines, and he kept complaining the company credit card never had enough money to buy fuel. We found out a company employee was stealing money from the bank account and didn't leave enough to pay for fuel. One day he came to Arizona to meet everyone, and when I met him, I told him to his face "I don't trust you." He tried to get me fired, but John De Palma wouldn't hear of it.

Being short on funds for fuel, went on for a year and a half, and finally, iJet went broke because of all the money was stolen from the company.

Right after IJet declared bankruptcy, I got a call from a company in Riyadh, Saudi Arabia asking I come back and help them start up an airline operating Boeing 737-300s. I agreed, packed my things and flew to London England and met everyone associated with this new company called Sama Airlines. The interview lasted an hour, and their last question was what my worst trait is. I thought and thought and my answer was, I should never have left my first wife. Everyone laughed, and the interview was over.

Conspiracy

They asked if I could go to Casablanca, Morocco before going to Saudi Arabia to review the simulator training at a local airline. I declined and said I would go after getting established in Riyadh. They agreed, and I left London for Riyadh the following morning.

Sama means "Morning Star" in Arabic. After getting settled in at my new residence, I traveled with UAE Airlines to Casablanca, Morocco to observe simulator training for the new flight crews.

We were in a taxi cab going to a restaurant for dinner, and the cab driver pointed to a building and said this was Rick's place in the movie Casablanca with Humphrey Bogart. My friend Greg Knapp worked for a movie publishing company in Burbank, California he told me the movie, had been filmed locally in Burbank. I didn't have the heart to tell this cab driver anything further.

We went to a local restaurant for dinner which had at the entrance a glass tank filled with water and about six lobsters with claws. To my knowledge, any lobsters with claws had to have come from the Boston area. I ordered one of the largest lobsters in the tank. When I got the check, it cost me one hundred and twenty dollars. The lobster was good, but not that good. It was huge, and I couldn't eat it all, so I shared it with the others at the table.

I hated Riyadh because there was nothing to do there except eat make homemade wine and drink. Working for the company was a little nerve-racking because my boss and his boss and CEO knew nothing about running an airline. In three years the airline shut down, and I believe it was because of the lack of proper management.

I went back home after my contract of two years was up, and my replacement was even worse than upper management. They hired someone local before I left, so I worked with this guy for a few weeks. He attempted to put in new procedures that meant nothing and would do nothing, and he asked me why I fought him on this. I told him because you don't know what you're doing. He walked away from me in a huff but didn't push the issue. From what I understand from Captain Kamal El Gamdi, I was right, he didn't know what he was doing and limited Kamal's activities to where he couldn't do his job of safety surveillance on the pilots.

I hired a flight attendant as one of my assistants, Mona Hawas, who was an outstanding employee and very dedicated to her job of evaluating Flight Attendant safety activities aboard Boeing 737 aircraft. Her cousin in Cairo is Zahi Hawas, curator of the ancient culture of Egyptian Antiquities, and who is watched on TV anytime there is a documentary featuring ancient Egyptian cultures. He is the individual you see with specks of gray hair throughout his black hair and speaks very good English. Mona is now a Flight Attendant for Saudi Arabian Airlines in Jeddah.

I returned to the US right after the first of January 2009. I sold my Harley Davidson Road King and bought a Harley Screaming Eagle Electro Glide painted Candy Apple Red. It was beautiful.

I was riding along Alma School Road going north to Greasewood Flats to have lunch with Steve Hedden when a BMW automobile came out of a side street from the right and hit me broadside. The last thing I remember is this gray car hitting me from the right. I woke up in the Ambulance going to the hospital. I wasn't wearing a helmet and so just the fact I was still alive amazed me. Both knees were severely damaged, and my doctor recommended replacing both of them as quickly as possible.

I waited almost a year before having the operation. I asked the doctor to replace both knees at the same time. The doctor was reluctant but eventually told me he would do it only if I stayed in the hospital for two weeks for physical therapy. I agreed but was released early after only a week and a half. The next day after leaving the hospital, I walked into the doctor's office without any assistance from crutches or a walker. The doctor told me no one has ever done this after having both knees replaced. I heal fast.

When I was on the operating table, I asked my doctor if he could put spacers on my new knees. I wanted to increase my height to six foot four inches. We all laughed. When I woke in my room, I looked at the personal information chart on the wall, and it read my height six foot four inches tall.

It took almost a full year for me to gain back the full usage of my legs, and today I try to walk three and a half miles every other day.

After recouping from the operation, I went to a party at a friend's house who was a CPA. He asked if he could borrow twenty-five thousand dollars from me for sixty days. I said OK but only if I can get my money back in the sixty days, and I gave him a check made out in his name for the twenty- five thousand dollars.

Three days later I received a package with documents transferring fifty-one percent of his company, "Entry Designs" into my name. I called him and told him I didn't want any portion of his company. He told me this was collateral for the loan. I insisted I didn't want any portion of his company and he went upset and yelled, I'm only trying to protect you. His statement turned out to be one big lie.

Since he was a friend and a CPA I figured he knew what he was doing, so I signed the papers. Unbeknownst to me, he had placed another document under the contract so that when I signed the contract, my signature also appeared on another document, supposedly assigning the $25,000 to the purchase his company Entry Designs.

Since I was still recuperating from my surgery, he asked if I could take over the accounting for the company. I told him my expertise was in aviation not accounting. He pleaded and since I wasn't doing anything other than physical therapy I said OK.

He then asked if I could pay a couple of suppliers the money Entry Designs owed. The reason was to preserve their sixty-day payment plan and that I would get my money back as soon as the payment arrived. One more of his lies. Two additional times a supplier asked for more money and twice again, and I never received any money in return.

I got suspicious and contacted the bank from which I wrote the original check. I couldn't believe what I saw. The remarks section of the check had the notation, "Purchased 51% of…" and it was not my handwriting.

I called him and told him what I had discovered and that I was going to come down there and rip his head off. He panicked and told me he would pay every cent back to me with six percent interest. If I called the police, which I considered, he would be arrested and charged with fraud, and I wouldn't get my money back.

I received a contract from him paying me $500 a month which including six percent interest just like he promised. Payments were received eleven months and then all of a sudden he liquidated Entry Designs and stopped paying me the money he owed.

I immediately contacted an attorney and explained the situation. An agreement to file a lawsuit against him reached and signed into existence. I paid thousands of dollars to hire this attorney, and it wasn't until about seven months later I found out through a mutual friend that the attorney never filed any lawsuit. When I confronted the attorney, more lies were given me about the reasoning, and none of them contained in the agreement we signed. It became obvious that once again I was at fault for trusting an attorney. I not only had a dishonest CPA, but I also hired a dishonest attorney.

I filed a complaint with the Arizona Bar Association. We agreed to meet and discuss our options, but when we met, it became obvious he was taking the side of the attorney I hired and made all kinds of excuses for not filing the lawsuit. No discussion of the signed agreement and its content discussed.

It was the same reaction when I went to the State's Accountancy Department, with the complaint of this CPA not being ethical in his activities. Arizona State Accountancy Department certifies every CPA in the state. Every year the CPA is required to complete an Ethical Training Course put on by the Accountancy Department. The Accountancy Department investigated my complaint and sent their investigator to talk to me. I explained everything, and he agreed this was a classic example of fraud. However, I was notified by the Accountancy Department they were going to close this investigation with no action but did not give a reason for their decision.

Chapter 10

STONE AIR AVIATION

Lynn Babcock bought a Falcon 10 which is a twin-engine jet French Built airplane that was the second fastest corporate airplane in existence. Lynn hired me as captain, and I received training from a retired Delta Airlines Captain. We trained while flying Lynn to Chicago for a week. While training, we hopped from Chicago to Kansas City, Missouri to Fort Worth, Texas, where I took my flight test for the issuance of a Falcon rating. Afterward, we flew back to Chicago to pick up Lynn and returned to Scottsdale. My instructor caught an airline out of Phoenix Sky Harbor Airport and went back home to Atlanta, Georgia.

Lynn hired a certified mechanic as co-pilot on the Falcon. Both he and Lynn went to Dallas, Texas to SimuFlite and received certification training in the Falcon 10. Lynn did not get rated, but the mechanic did.

Meanwhile, I was proceeding with FAA certification for obtaining authorization to carry passengers and cargo for Lynn's company called Stone Air Aviation. The final certification phase was for us to demonstrate to the FAA, our ability to conduct flight operations internationally. The destination given us by the FAA was St. Thomas, an island in the U.S. Virgin Island chain in the Caribbean.

Aboard the aircraft were three FAA Inspectors, one Operations Inspector, an Airworthiness Inspector (mechanic) and an Avionics Inspector. The flight took us to Florida where we refueled and then direct to St. Thomas Island. At night over the ocean when no lights are visible, it is totally black, and you can't see anything. It is like looking into a black hole, and I heard one of the FAA inspectors comment that I hope Al knew where he was going. I laughed.

We landed at St. Thomas and met a friend of mine Ray Morgan who happened to be there at the same time we arrived. Naturally, this meeting was prearranged by me before the flight even left Scottsdale.

Ray drove us to our hotel and the next day went sightseeing. I had been on St. Thomas Island before so we toured the island for the benefit of the FAA Inspectors. They all were great to be around, and it was obvious the vendetta was over. It was a big relief.

One of the FAA Inspectors bought a wig that had curls hanging down like a Jamaican native and started directing traffic at a busy intersection. He was hilarious.

On our flight back to Scottsdale we stopped at West Palm Beach, Florida for fuel. Due to severe weather just west of our location, we stayed the night. The FAA inspectors liked my decision and made the comment they were impressed with the way both co-pilot I and worked together. The following day we returned to Scottsdale and received our certification from the FAA.

Our next trip was to St. Cloud, Minnesota delivering three passengers to their home. After landing, I shut one engine off to conserve fuel while we taxied to the ramp. The Falcon airplane has two hydraulic systems, one on each engine. The normal brake system is on the right engine, and when it is shut down, you must turn on an electric hydraulic pump to keep the brakes working.

After landing, I shut the right engine down and turned on the electric hydraulic pump. I then concentrated on taxiing the airplane to the ramp. Undetected by me the co-pilot completes the after landing checklist but mistakenly shuts the electric hydraulic pump off taking away the hydraulic pressure to the main brakes.

I continued taxiing to the parking area, and an individual put up his hands signifying he will guide me to a parking spot. It became obvious he was guiding me to a parking spot in front of a plate glass window. As I approached, the ramp person held up his two arms crossing them at the wrist signifying me to stop. I applied brakes but found no one home, and the airplane continued toward the plate glass window. I immediately applied full left nose wheel steering and applied emergency brakes to stop the airplane just before the right wing went into the plate glass window. Another Whew!

Conspiracy

I told my co-pilot to announce anytime he turns anything on or off so that I know what is happening. Making these announcements is a normal procedure anytime he completes a checklist. He apologized and agreed.

I had the airplane refueled after arrangements were made to transport our passengers to their home. We started engines and taxied the airplane to the active runway and departed St. Cloud for Scottsdale. During the climb and passing through twenty-six thousand feet I felt hypoxia. Hypoxia is a phenomenon where the body begins to go to sleep due to the lack of oxygen, and if not corrected, eventually death will follow. I looked at the pressurization gauge, and it read zero. I told the co-pilot to get his oxygen mask on immediately and that we lost pressurization. I declared an emergency with Air Traffic Control and told them of our problem and that we needed to get a lower altitude. The controller cleared other traffic out of our way, and we descended to a lower altitude below ten thousand feet. Ten thousand feet is the maximum altitude a person can survive without oxygen.

Fortunately, I had received training in an Altitude Chamber to identify my symptoms for detecting hypoxia. My first thought was that golfer, Payne Stewart when he was traveling in a Learjet Airplane and lost pressurization and no one detected it. They all eventually died, and the airplane continued to fly because the autopilot was on.

When Air Traffic Control couldn't contact the Learjet, they alerted Military Jets to fly alongside the Learjet to see what was wrong. They couldn't see in the airplane because the windows had frozen over. The airplane eventually ran out of fuel and crashed killing everyone on board, although they had probably died long before the crash. Researching pilot records of the flight crew, it was determined no one had ever received training in an altitude chamber.

My co-pilot told me he shut the pressurization down to see how much additional power was available from the engines during takeoff but forgot to turn it back on. When we arrived back in Scottsdale, I fired him because I didn't like him turning things on and off without telling me. My training in an Altitude Chamber saved our lives. Unless you have experienced how your body reacts to the loss of oxygen, you might not be able to detect a pressurization failure.

Normally, a warning horn sounds when the cabin altitude reaches fourteen thousand feet. However, when the ground crew in Fort Worth had removed the interior for new upholstery, they disconnected the emergency altitude warning system. During preflight, a test button is pressed to test the emergency altitude system. Every time it was tested, results were that is was functional.

Hal Price, hearing of my incident with my co-pilot, asked that I hire a friend of his who is retiring from America West Airlines to replace him. I said OK and hired him. Hiring him was another one of my big mistakes.

I trained him, got him his rating in the Falcon 10 and the FAA Ok'd him to fly as co-pilot during air carrier operations. By this time, Lynn had sold half interest in the Falcon 10 to an individual named Gene. My co-pilot began talking to him about firing me and hiring him as captain. He based this decision on the fact he was an airline pilot and knew more than I did. Gene didn't know any better and not being there to defend myself; I was eventually laid off.

Bob wasn't that good a pilot, and I would constantly correct him when operating the GPS navigation. During our recurrent training at SimuFlite, he constantly exceeded the aircraft limitations and failed the check ride and the recurrent training. Of course, Gene never knew this, and when I told him, I revealed Fred's shortcomings but Gene had already made his decision, and he was not about to admit he made a mistake. Bob was unqualified to fly the airplane as captain but did so anyway.

So, I was unemployed. Eventually, Gene found out what kind of person Bob was, and he fired him on the spot. Of course, it was too late for me.

When I told Hal Price about this, of course, Hal wouldn't admit he should never have asked me to hire Bob. Once again, doing a friend a favor I got screwed.

Chapter 11

ARIZONA DEPARTMENT OF TRANSPORTATION

Since I could no longer fly because of losing my medical in that motorcycle accident, I began driving my Dodge RAM 3500 pickup truck for Synergy RV Transport in Goshen, Indiana delivering 5th wheel RVs and vacation trailers all over the US and Canada. I've seen this entire country and much of Canada flying a jet at 41,000 feet. Now I'm seeing it from the ground, and I enjoyed it.

I've been from coast to coast in the US and from Vancouver Island to Nova Scotia in Canada and much of in between. On the 6th of September 2015, I entered the Port of Entry to Arizona at Granada which is thirteen miles west of the New Mexico Border on Interstate Highway I-40. My last stop had been in Amarillo, Texas where I became sick and unable to drive. This condition caused me to stay in Amarillo for three hours. Finally, I felt I could drive safely to my planned overnight stop at Granada, Arizona.

I also had to purchase a permit to deliver this commercial trailer to Yuma, Arizona. As I stopped an Arizona Department of Transportation (DOT) Officer came up to me and asked that I reposition my truck so that he could inspect everything. I told him I had reached my limits for driving and that I had been sick in Amarillo for three hours before continuing.

In my mind, I would have no problem with the inspection because I felt everything was in order except for the last entry in my driver's log. As I gave him my driver's license, he commented on my motorcycle endorsement. I thought he wanted this inspection to be formal but friendly. So, I told him about riding motorcycles with Osama Bin Laden's brother Yassah when I lived in Saudi Arabia. He then told me I should have had this last entry completed before entering the Port of Entry. I explained that would be dangerous to fire up my computer and attempt to make entries with the mouse while driving. I explained that would be more dangerous than driving and texting on my phone.

He then demanded I not complete anything other than the one last entry and nothing else. He then threatened me that if I didn't sign these logs when I printed them out, he would file a violation against me. So I signed all the logs including the one he wouldn't allow me to complete. I felt it was like signing a blank check.

Based upon the incomplete log, he filed a violation against me for exceeding driving limitations of eleven hours, exceeding duty time limitations of fourteen hours, and falsifying two logbook pages. One for having two gas slips on the same day, and the other because of the date. This DOT officer contrived these infractions, and It was obvious talking to this officer that he had made his mind up and there was no altering it.

My plan when leaving Amarillo was to be at my driving time of eleven hours when I arrived at the Port of Entry in Granada, Arizona. I had expected to purchase a permit and stay overnight at the Port of Entry and to continue to my destination of Yuma the following morning. I was four minutes early arriving at the Port of Entry.

Someone in the DOT and I'm assuming it was the officer conducting the inspection, called Synergy RV and told the company I violated certain DOT regulations. Synergy immediately notified me that I was suspended but of course, after I delivered the RV to Yuma, basically taking away my livelihood. The DOT, however, didn't suspend my driving because of the allegations. Then why did Synergy suspend me when the DOT didn't?

Even though each state has their own Department of Transportation, they all use the government's DOT regulations to keep the same requirements throughout the country. The DOT for the State of Indiana conducts inspections at Synergy RV Transport evaluating the accuracy of drivers' logs throughout the year and grades the companies accordingly. Indiana's DOT had given Synergy RV Transport a grade of 86, which is considered very bad.

Conspiracy

Suspending a driver and taking away their livelihood, whether the charges are legitimate or not puts a hindrance on the driver because now all pay stops and they are unable to pay their normal bills. This suspension stays in effect until this violation is taken care of by paying the fine whether or not the charges are legitimate. What can the driver do? He/she pays the fine just to get back on the payroll. I couldn't do that because; being an ex-cop myself made me furious when I see dishonest Public Servants.

I purchased a professional CD containing all Department of Transportation (DOT) rules and regulations. In a telephone conversation with this company, I described my experiences with Arizona's DOT. This individual told me he has heard this same scenario happening to drivers all over the country and he felt this was a conspiracy between the trucking companies and each state's DOT. He felt this was just a way for the states to increase their revenues at the expense of the truckers.

Had I known the DOT regulations when confronted with this officer I could have countered his accusations by quoting the applicable regulations. However, at the time I didn't know them at all. Now I do.

Being an ex-law enforcement officer myself, I have a real problem with dishonest cops, and this is a classic example of dishonesty in the DOT.

I talked to the owner of Synergy and asked him why was I suspended, and he said we have to talk. I couldn't believe him, and I asked, "Talk about what?" He said about driving. I told him if I drove all the way to Goshen to talk, there was no guarantee that he would take me off restriction if these charges were still in place. He agreed that could happen. So I resigned. I didn't want to work for someone who betrayed his drivers to advance the company in the eyes of the DOT. So now all I have to worry about is how to pay bills.

The DOT Regulations state that during an inspection by a DOT officer, the drivers must be allowed to complete two full days of driving logs. I was only allowed to complete one line on the log and not the entire log. That left the driving and duty times blank leaving it open for anyone to make false entries. I carried a book of DOT regulations in my truck, and after reading the regulation concerning the driver being allowed to finish two days of logs as opposed to this officer's limit of just one entry, I completed the log for that day. After making the final entries, the log showed no violations existed.

When I completed the log for the day I got inspected, it would have shown no violation existed because I was under the limits for driving and duty time. The day I was supposed to falsify a log was a trip I made when the truck did not meet the DOT's definition of a commercial vehicle. The log had a notation in the log that it was a "Dead Head" trip, meaning I did not receive remuneration for this trip. I was empty for I was returning to Goshen, Indiana to pick up another RV for delivery.

The other falsification showed the same notation "Deadhead." Again, means I was heading back to Goshen, Indiana after delivering the RV to its destination. This trip was made with no trailer and consequently does not fall under the definition of Commercial Vehicle under the regulations of the DOT because no remuneration was received. The cost of this trip was by me out of my pocket.

The summons I received from the DOT Officer demanded I appear on October 14th, 2015 at the Justice Court, County of Apache, State of Arizona, Puerco District, Sanders, Arizona, two hundred and eleven miles from my home in Scottsdale, Arizona.

On October 14th I drove four hours and some two hundred and eleven miles one way to appear in Sanders court. I started explaining to the judge what had taken place and he immediately stopped me and said I would have to return because the County Attorney was not present. My feelings were why was I even there without the County Attorney present? At the request of the judge, I pleaded not guilty and was ordered to return on November 19th, 2015.

I returned on November 19th, 2015, and the County Attorney was present. Before the judge's arrival in the Court Room, the County Attorney offered me a Plea Bargain telling me that if I pleaded guilty, he would impose a reduced fine and have me attend driver's school. I refused his offer not telling him my reasoning.

When Judge Yellowhorse convened the court, the County Attorney made a motion to dismiss the case without prejudice because the accusing officer didn't show up in court nor did he submit a report of the so-called violation. When the accusing officer failed to show up in court, it violated my Constitutional rights by not allowing me to face my accuser (Amendment VI of the Constitution). The definition of "Without Prejudice" permits the DOT to refile the complaint if they receive additional evidence to justify the reopening.

I wrote the Director of the Department of Transportation a letter asking they reimburse my expenses for operating my truck on two round trips from my residence to the Sanders Court House and return since the court placed the fault for these infractions on the DOT. My letter to the DOT ended up with a Tim Lane, Head of Security for action. Mr. Lane's letter stated they don't reimburse anyone for attending court even if the case was the fault of the DOT. A few days later I received another summons reopening the original case and for me to again report to Sanders Court on the 16th of January 2016, in the Sanders Court. It was obvious this was retaliation against me for the letter I wrote to the DOT asking for my truck expenses.

 I am not an attorney and don't profess to be, but I don't have the funds to hire legal help. It would be cheaper to pay the fine than hire an attorney, and I can't even do that. So, I have to fight it. That DOT Officer fabricated these charges because I didn't know the regulations good enough to dispute his allegations when he first presented them to me on September 7th, 2015. The DOT hid the officer who made these charges against me and did not allow him to show up in court. They even tried to use another officer who had nothing to do with issuing me the ticket. Again, this would have violated my constitutional rights using someone who was not my original accuser.

 I returned to court, and the judge rendered his decision finding me at fault on two charges. Exceeding duty time and driving time. He also found me guilty of falsifying a driver's log. The other violations of regulations he found me not guilty. When I confronted the judge on finding me guilty of falsification on one charge and not guilty on another, I questioned him because both charges were for the same violation. He then told me I would have to appeal his decision.

 I filed my appeal and had to return to court. I again entered my plea of "not guilty" with the understanding I had to return to court at a later date for the appeal process.

 I returned, and there was a new judge on the bench, and he presided over this appeal. The prosecuting attorney asked for dismissal of the charge exceeding duty times and driving times.

 The prosecuting attorney presented the same evidence as last time, and I didn't object to any of it. When it came time for me to present my case, I showed the court the numerous pieces of evidence presented by the county attorney showed the truck did not qualify as a commercial truck under the DOT definition. The definition of commercial truck was one that resulted in the operation receiving remuneration of some sort. I did not receive any remuneration, and the driver's log showed this because there was a statement entered in place of a Bill of Laden number "Deadhead." Deadhead means no remuneration of any kind was received. Also, on my behalf the county attorney referred to this statement and that if it were a commercial vehicle, there would have been a Bill of Laden number entered. I did not have a Bill of Laden number entered anywhere on the log.

However, even with this evidence entered into the court's record, the judge disregarded the DOT regulations and declared it was a commercial truck and fined me $500.45.

I feel this shows the dishonesty of Arizona's court system and that it will do anything to protect the Department of Transportation. I'm sure this dishonesty goes deep within the management of the Department of Transportation, encouraging their officers to fabricate erroneous charges to gain unjustified fines. Conferring with J & J Keller publisher of commercial DOT regulations, I feel this falsification of charges is continuing throughout the truck industry, and its success is backed and supported by dishonest Judges.

EPILOGUE

I believe this book has shown the dishonesty existing throughout our government system, both National and State organizations. It shows the state's attempts at increasing revenues by fabricating charges against truckers, and if anyone challenges them, the court system is ready to step in and protect the DOTs dishonesty.

My Niece Judy, (my Brother's oldest daughter) lost her means of support at a time when she was eight months pregnant with her youngest son Michael. She was destitute and wasn't able to support herself, so when she heard that the FAA Air Traffic Control Tower Operator (ATC) at the Van Nuys Airport demanded a pilot execute a hazardous maneuver, not in the ATC Handbook and the aircraft crashed killing its occupants, she rightfully demanded restitution.

Since I taught her husband Vince how to fly, and was his instructor for his instrument rating, I was subpoenaed to appear in court by Judy's attorney. Even though I had no choice but to testify because of the subpoena, U.S. Attorney's, would call me at home, in my office and threaten my career if I testified. FAA orders allow me to testify in court if it involved immediate family, so long as I didn't provide an opinion, suggestion or any suppositions. I could testify only to facts about Vince's ability to fly as a pilot. Even though according to FAA Orders, I was allowed to testify, I was threatened and coerced not to testify. I received a subpoena, so I didn't have a choice, but that didn't matter.

Just before the court date, I received numerous telephone calls from the FAA's legal department threatening that if I testified; I could either lose my job, or be transferred, or assigned to anywhere in the FAA. One day before the trial I was in Judy's attorney's office when I received a threatening telephone call from the Western-Pacific Regional Attorney's Office, Mr. Dewitt Lawrson. I reminded him of where I was, and if he didn't stop threatening me, I would file an action against him.

The judge in the trial knew and understood what my limitations were when testifying and made a statement to the court regarding this limitation. Neither attorney made any objections. I testified being careful not to compromise my position. But this didn't matter to the FAA. The trial ended, and the judge over a year before rendering his decision. He found the FAA to be one hundred percent at fault. The decision came down that Judy was to be awarded one million dollars for Vince's death.

During the initial stages of the FAA's Vendetta, I lost my Pilot Examiner Designation. The Security Division of the Regional Office initiated an investigation accusing me of a conflict of interest. Witnesses all of a sudden appeared accusing me of forcing them to buy one of the computers I was building. The fact was that I never sold any of these complainers a computer which rendered their investigation mute.

The Flight Standards Division Manager, my boss's boss in the Regional Office sent me a letter demanding me to move back to the Regional Office in Los Angeles because of my helicopter experience. It was a joke because there were five other individuals in the Los Angeles with extensive helicopter experience. He didn't have to move me other than for vindictive purposes. However, this did provide me with enough qualifications to retire early.

In the Division Managers letter, he gave me sixty days in which to sell my house, pack and move to Los Angeles. Right after receiving this letter, I received word that since I was qualified and current in a Sabreliner, I was to travel to Tokyo, Japan and do an audit of the International Flight Inspection Operations. The length of time it would take me to complete this audit, I might not have enough time to sell my home and report to Los Angeles on the date indicated in the letter thereby giving them the excuse they wanted to fire me.

So I retired figuring it would remove me from any further retaliation from individuals in the FAA. How little did I know?

The FAA felt perfectly safe regardless of whether or not they were at fault. I even wrote Senator McCain filing a complaint of what the FAA was doing to me, and he did nothing. I complained that the FAA did not have justification not to renew my Pilot Examiner Designation. They were counting on a conflict of interest when the investigator found nothing

Conspiracy

and his investigation went mute. Even the Ninth Circuit Court of Appeals recognized the fact Tom Accardi, AFS-1 had a vendetta against me.

When employed as a pilot with Eastern Airlines, I was home in Scottsdale, Arizona on my days off; and I was approached by one of FAA's regional councils to testify as an expert witness in a case they had pending against a Phoenix's Channel Twelve helicopter pilot. The FAA attorney stated I would receive $500 a day for case preparation, consulting and testifying. I accepted his terms and contacted Eastern Airlines and received their approval for me staying in Scottsdale.

At the FAAs requested I submitted my bill for services rendered and received a letter from Mr. DeWitt Lawson, head of the FAA attorneys and who has been at the heart of the vendetta. He denied the $500.00 a day and lowered the amount owed me to $200.00 stating this as a limit according to agency limit. After researching his statement, I found documented evidence that his statement was a lie but I elected not to pursue it.

I have collected and saved all documents showing the FAA vendetta, and Figure 26 is a photograph of the complete collection of all letters, statements, reports, and evidence showing the FAA's vendetta against me since my brother's death.

FIGURE 30 – ACCUMULATION OF DOCUMENTS PROVING FAA'S VENDETTA

Printed in the USA
CPSIA information can be obtained
at www.ICGtesting.com
LVHW050849111123
763265LV00121B/2139